International Conference on Advanced Materials and Technology for Sustainable Life (ICAMTSL 2019)

IOP Conference Series: Materials Science and Engineering
Volume 587

Shanghai, China
28 – 30 June 2019

ISBN: 978-1-5108-9460-0
ISSN: 1757-8981

Printed from e-media with permission by:

Curran Associates, Inc.
57 Morehouse Lane
Red Hook, NY 12571

Some format issues inherent in the e-media version may also appear in this print version.

Institute of Physics
Dirac House, Temple Back
Bristol BS1 6BE UK

Phone: 44 1 17 929 7481
Fax: 44 1 17 920 0979

techtracking@iop.org

Additional copies of this publication are available from:

Curran Associates, Inc.
57 Morehouse Lane
Red Hook, NY 12571 USA
Phone: 845-758-0400
Fax: 845-758-2633
Email: curran@proceedings.com
Web: www.proceedings.com

Table of contents

Volume 587

2019 International Conference on Advanced Materials and Technology for Sustainable Life

28–30 June 2019, Shanghai, China

Accepted papers received: 8 July 2019
Published online: 29 August 2019

Preface

Papers

Application of Advanced Materials in Sustainable Development

IOP Conf. Series: Materials Science and Engineering **587** (2019) 011001 doi:10.1088/1757-899X/587/1/011001

Preface

The organizing committee of ICAMTSL 2019 is proud to present the proceedings of the 2019 International Conference on Advanced Materials and Technology for Sustainable Life (ICAMTSL 2019: http://www.icamtsl.net/), held in Shanghai, China from 28 to 30 June 2019. We warmly welcomed prospective authors to submit new research papers to ICAMTSL 2019 and share their valuable experience with scientists and scholars around the world.

The conference is an international forum for the presentation of technological advances and research results in the fields of sustainable aspects of materials science and technology such as fundamentals, applications and engineering of biomaterials, nanomaterials and composite materials. The conference will bring together leading researchers, engineers and scientists in the domain of interest from around the world.

ICAMTSL 2019 received more than 50 manuscripts, and 17 submissions have been accepted by our reviewers. By submitting a paper to ICAMTSL 2019, the authors understood and agreed that papers would undergo a rigorous peer-review process. Manuscripts were reviewed by at least two independent, qualified experts in the field selected by the conference committee, who took detailed comments and the authors would submit a revised version in which the feedback was taken into consideration. All papers were reviewed using a double-blind review process: authors declared their names and affiliations in the manuscript for the reviewers to see, but reviewers did not know each other's identities, nor did the authors receive information about who had reviewed their manuscript.

The committees of ICAMTSL 2019 invested great efforts in reviewing the papers submitted to the conference and organizing the sessions to enable the participants to gain maximum benefit. We sincerely appreciate your participation and do believe our joint efforts would make our conference indeed fruitful and successful.

With our warmest regards,

Conference Organizing Chair

Shanghai, China

IOP Conf. Series: Materials Science and Engineering **587** (2019) 011002 doi:10.1088/1757-899X/587/1/011002

Committees

Conference Chairs

Prof. Dengguang Yu, University of Shanghai for Science and Technology, China

Prof. Gareth R Williams, University College London, UK

Program Chairs

Prof. Xiaonan Xiao, Xiamen University, China

Prof. Bo Li, Xi'an University of Posts & Telecommunications, China

Program Committee

Prof. Xueyan Lin, Beijing University of Posts and Telecommunications, China

Prof. Ghenadii Korotcenkov, Moldova State University, Moldova

Prof. Fei Li, Chengdu University of Technology, China

Prof. Limei Wang, Beijing University of Technology

Prof. Joseph Yu-shek CHENG,City University of Hong Kong, Hong Kong

Prof. Xifu Xu, Shandong University, China

Prof. Qingao Yue, Dalian Maritime University, China

Prof. Jun Lin, Xi'an Jiaotong University, China

Prof. Cheng-Ping Chang, National University of Tainan, Taiwan

Prof. Jiafeng Liu, The School of History and Culture, Shandong University, China

Prof. Fangwen Zhu, Shanghai University, China

Prof. Shuangmei Liu, Nanchang University, China

Prof. Deyun Mo, Lingnan Normal University, China

Prof. Elnur Mekhdiev, Financial University under the Government of the Russian Federation, Moscow, Russian

Prof. Mengxi Xu, Nanjing Institute of Technology, China

IOP Conf. Series: Materials Science and Engineering **587** (2019) 011003 doi:10.1088/1757-899X/587/1/011003

Peer review statement

All papers published in this volume of *IOP Conference Series: Materials Science and Engineering* have been peer reviewed through processes administered by the proceedings Editors. Reviews were conducted by expert referees to the professional and scientific standards expected of a proceedings journal published by IOP Publishing.

IOP Conf. Series: Materials Science and Engineering **587** (2019) 012001 doi:10.1088/1757-899X/587/1/012001

Recyclability of plastic waste extracted from electrical devices

I Turku[1*] and T Kärki[1]

[1]LUT University, Yliopistonkatu 34, 53850 Lappeenranta, Finland

*Corresponding author and e-mail: Irina.Turku@lut.fi

Abstract. The mechanical recyclability of acrylonitrile butadiene styrene, high impact polystyrene, polycarbonate and polypropylene extracted from electric equipment waste was studied. The secondary raw material was tested on a broad range of working characteristics, such as melt flow index, thermal properties as well as tensile properties of injection molded samples were detected. Parameters were compared with parameters of unprocessed counterparts presented in the literature.

1. Introduction

Waste from electrical and electronic equipment (WEEE) grows up very fast that requires optimization of its utilization process. About 17 wt% of total WEEE stream is plastic components. The most abundant plastics in WEEE are acrylonitrile butadiene styrene (ABS), high impact polystyrene (HIPS), polypropylene (PP), polycarbonate (PC) as well as polyvinylchloride, polybutylene terephthalate and polyamide, which counts up to 70 wt% of total plastic collected [1]. According to statistics, only 25% of e-waste plastics is recycled [2] that is lower than the average rate of post-consumer plastic utilization. Thus, from 27.1 Mio of plastic waste generated in Europe in 2016, about 31.1 % was recycled, 27.3 % landfilled and 41.6 % went to incineration [3]. The main challenges for EEE plastic utilization are the complexity of the different types of plastic used in EEE products, black colored materials, lack technics for separating all plastic types into individual streams and limiting understanding of the effect of the mixing of different plastics on the physical properties of the recycled materials [4]. In addition, the presence of halogen-containing fire retardants and heavy metals negatively influence on the recycling rate.

Carbon-based reinforcing fillers, namely carbon fibers, graphite, carbon nanotubes, carbon black (CB), represent a great interest in the composites manufacturing due to their multifunctional properties including high strength-to-weight ratio, high electrical and thermal conductivity, and fire resistance among others. Many useful properties of CB like variety in structure, low cost and capability of CB particles to form strong attachments to elastomer molecules without using expensive compatibilizer make CB very important and most widely used filler in rubber and plastic industry [5]. Concerning polymer waste recycling it was shown that incorporation of 3-10 % of CB resulted in the improved tensile strength of a recycled PP, ABS and PS up to 12-23 %, depending on the plastic source [6]. The tensile properties of PP and recycled natural rubber glove blend was improved by loading of CB [7]. CB had a positive effect on the tensile and impact strength properties of unprocessed PP and ABS blend [8]. Comparative analysis of compression and injection molded PP/CB nanocomposite showed that injection molded samples have a superior tensile property than compression molded [9].

The aim of this study was to estimate the recyclability of ABS, HIPS, PC and PP extracted from EEE waste. Pieces of plastic were reduced until microsize and then were used for melt flow index

(MFI) and thermal analysis as well as samples for tensile properties were injection molded. The influence of carbon black filler on the tensile properties was also studied. In addition, elemental analysis of the waste was performed by using energy-dispersive X-ray spectroscopy (EDS).

2. Materials and methods

2.1. Plastic source

There are four types of WEEE plastic, ABS, HIPS, PC and PP were used in this study. The materials were obtained from the post-consumed electronic devices, Figure 1. The coffee machines plastic parts were sources of PP, the refrigerator plastic parts were the source of PC, different plastic parts of computer were ABS and HIPS sources. Plastic pieces were additionally analyzed with FTIR (data not shown). Carbon black was Raven ®1170 grade, mean particle size 21 nm and surface are 107 m^2/g (IMCD Finland Oy).

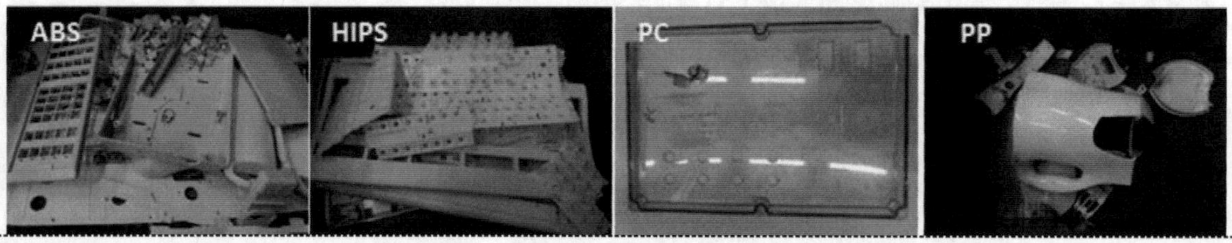

Figure 1. Sources of plastic waste.

2.2. Samples manufacturing

The plastic fragments were reduced to 2 cm - sized flakes by low-speed granulator Shini SG-1635N equipped with a 5 mm sieve. The samples for testing were prepared in a Boy 30 injection-molding machine. The parameters used for injection molding were the following parameters: T_1=160 °C and T_2=170 °C for ABS, PS and PP, respectively; and $T_{1,2}$=210 °C for PC; injection pressure 8.0–9.1 MPa and injection time 3 s. In order to improve nanofiller dispersion, two-step injection-molding procedure was applied, e.g. samples were crushed after the first injection molding and molded again.

2.3. Testing

Melt flow index (MFI) was measured by using Dynisco LMI 5000 (Dynisco, Franklin, MA, USA) in accordance with standard EN ISO 1133-1.

The surface morphology of the samples was studied with an SEM, Hitachi SU3500 (Chiyoda, Tokio, Japan). Surfaces were observed directly after processing. Elemental analyses of the blends and single polymers were performed with EDS X-ray spectroscopy (EDS) (Thermo Scientific, Waltham, MA, USA).

The tensile properties test was performed according to EN-310 standard on a Zwick Z020 machine (Ulm, Germany). The cross-head speed was 2 mm/min for modulus testing and 50 mm/min for the other measurements. The gauge length was 25 mm. The test samples conditioned according to the above standard. Tests were carried out with 10 sample replicates.

Thermal analysis was performed by mean a thermogravimetric analysis (TGA) with a linear temperature increase (Netzsch DSC TGA 204 F1 Phoenix®). The sample, approx. 10 mg of the specimen was heated from 29 °C to 900 °C at a rate of 10 °C/min under a helium atmosphere of 40 mL/min at a constant flow rate.

Table 1. Elemental composition of ABS, PS, PC and PP, wt%; standard deviation is given in parenthesis

Element	C	O	Mg	Al	Si	Ti	Ca	S	Cl	P	N	K	Na	Cu	Mo
ABS	97.1	<0.1	0.01	0.3	0.04	1.1	-	0.06	<0.03	0.1	1(4)	<0.02	<0.01	-	0.05
	(0.1)		(0.01)	(0.2)	(0.02)	(0.2)		(0.05)		(0.03)					(0.08)
HIPS	99	-	-	0.2	0.04	0.6	0.02	-	-	-	-	-	-	<0.01	-
	(0.2)			(0.2)	(0.01)	(0.05)	(0.08)								
PC	83	16	-	0.4	0.04	-	<0.01	0.04	-	<0.01	-	<0.01	-	-	-
	(2)	(2)		(0.2)	(0.04)			(0.04)							
PP	99	-	0.06	0.25	0.17	1	0.1	<0.01	-	-	-	-	-	-	0.01
	(0.3)		(0.01)	(0.2)	(0.03)	(0.1)	(0.03)								(0.02)

3. Results and discussion

3.1. SEM-EDS analysis

Results of SEM-EDS analysis of secondary plastics are displayed in Table 1. As it can be seen recycled plastic contains different chemical elements which related to the additives used for a properties modification. Elements amount changes from traces to several percents. Distribution of detected elements is not uniform that is seen in the large standard deviation values, often exceeding the average amount of specified element. The main conclusion that can be made on the basis of the result, plastic waste does not contain toxic elements, such as heavy metals, Cd, Pb, and halogens, Br, Cl, except ABS which have traces of Cl. Halogen-containing compounds which can present in plastic can be attributed to fire retardants. Absence of Br among detected chemical elements can be explained by the fact that only very insignificant amount of plastic waste, 2.5% of total WEEE plastics, contain Br based fire retardants [10].

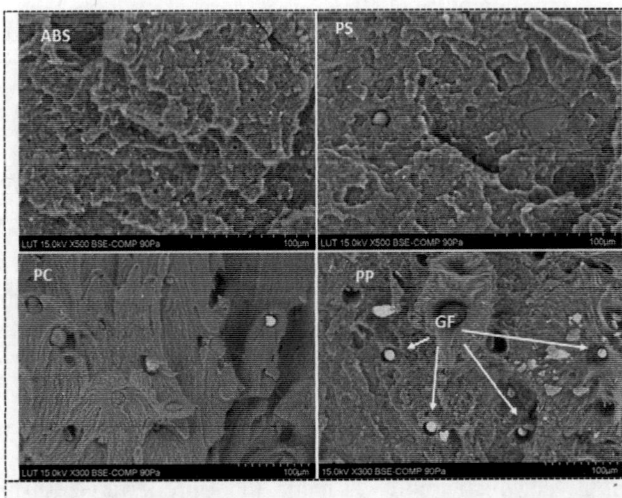

Figure 2. SEM micrograph of 3 wt% CB containing plastics fracture surfaces.

There are a broad range of other elements, such as Mg, Ti, Al, Na, K, Ca, Mo, Cu, S, P and O, originated from additives that are usually applied in plastic production or contaminants. Metal-containing fire-retardant $Mg(OH)_2$ and $Al(OH)_3$ are sources of Mg and Al, respectively. Phosphorus-containing flame retardants are widely used in the plastic industry [11]. The small amount of silicon, Si, which is a component of sand, can have originated from soil impurities. However, SEM analysis showed that PP waste contains glass fibers (GF), Figure 1, that can explain enhanced detected amount of silicon there. Nitrogen was detected in ABS due to it is a component of acrylonitrile. Oxygen is a component of PC that explain its 16 wt% detected amount. Oxygen was detected in trace amount in

ABS as well. Oxygen is a component of many chemicals including fire retardants, TiO_2 and other compounds often used in plastic manufacturing.

3.2. MFI analysis

Table 2 presents the measured MFI parameters for the recycled polymers and parameters found in the literature for virgin ones. For the comparative analysis of our samples and virgin ones, the database [12] was used where some range of MFI values is present. As it can be seen measured MFIs are fitted in the values range of the virgin counterparts, however, due to the absence of information/knowledge about recycled material applied grade it is difficult to make conclusions about morphological changes during plastic usage and recycling. In fact, MFI is sensitive to environmental impact and thermomechanical stress, e.g., during lifecycle and reprocessing. It was reported that the MFI of low-density polyethylene decreased from 2.31 g/10 min to 0.02 g/10 min after 100 extrusions [13].

3.3. Heat resistance and thermal stability

Experimental results of the thermal degradation of the polymers, mass losses and the corresponding differential thermogravimetry (DTG) curves, are shown in Figure 2. Pyrolysis of all studied polymers occurs in a one-step decomposition under the applied conditions. Polycarbonate is known as plastic which has good thermal stability and left a large fraction of char after burning [14,15]. As it is seen degradation started at about 430 °C with maximum weight loss at 483 °C and no mass loss happened after 650 °C. After the finish of pyrolysis about 20 wt% of char was left, that is significantly more compared to other tested plastics; 4.5, 2 and 3 wt% for ABS, PS, and PP, respectively. Other studied polymers started to degrade significantly early with peaks of mass loss at approx. 420, 430 and 454 °C for ABS, PS, and PP, respectively. Usually, plastics burn incompletely if some amount of inorganic additives, e.g. fire retardants, contaminant and/or reinforcing fillers are present in recycling materials.

Table 2. Parameters of extruded filaments for recycled samples and their virgin counterparts found in the literature [12].

Material/parameter	T_{extr}(°C)/mass	Density (g/cm^3)	MFI (g/10min)	MFI, virgin (g/10min)
ABS	220/10kg	1.06(0.06)	35.4(4)	2.1-36
HIPS	200/5kg	1.005(0.19)	4.9(1)	2.3-8.2
PC	300/1.2kg	0.996(0.16)	24.8(5)	1.5-24
PP	230/2.15kg	0.796(0.135)	15(5)	0.3-27

3.4. Mechanical properties. Influence of CB on the tensile properties

The tensile properties of the recycled plastics and their virgin counterparts reported in the literature are listed in Table 3. According to the results, the tensile properties of the recycled materials are fitted into the range of the virgin counterparts and some values even exceeded the maximal value. Thus, the tensile strength and modulus of PP were slightly higher than the highest values reported for virgin PP. This might be attributed to the presence of a reinforcing filler(s). As it was detected by SEM analysis, glass fibers were used in some PP based EEE, Figure 2.

CB is known as a reinforcing additive, which improves mechanical properties and durability of polymers. However, the properties of modified material will depend on the CB dispersion level. Incorporation of CB, 1-5 wt%, had no significant effect on the mechanical properties of the plastics, Table 3. The tensile properties started to increase only for PC when 5 wt% was loaded. In all other cases, the tensile strength was unchanged or decreased slightly with CB loading. The tensile modulus, however, increased in comparison to unfilled polymer almost for all cases. The increase in the modulus was due to the stiffness of CB is higher than polymer. Lohar and Jogi [8] studied the influence of CB on mechanical properties of PP/ABS blend and showed that the tensile properties of the blend increased to a maximum at 2.5 wt% of CB loading and then decreased with increased CB percentage. This properties droplet was due to CB particles aggregation. Uniform dispersion of CB resulted in the increased the tensile properties of the PP/CB nanocomposite with increasing CB content (1-30 %) [9]. Composites morphology inspection with SEM, Figure 2, showed that CB was aggregated that might negatively influence the performance of the material. The CB stay aggregated despite the double extrusion procedure applied for the processing of the samples.

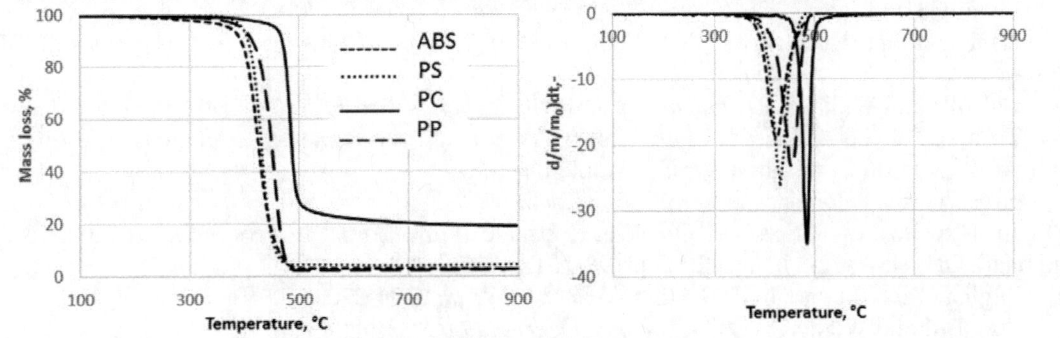

Figure 3. TGA analysis and differential thermogravimetry curves of plastics pyrolysis under a neutral atmosphere.

Table 3. Tensile properties of virgin [12] and recycled plastic filled with different amount of carbon black; standard deviation is given in parenthesis.

Plastic	Virgin		Reference		1 wt% CB		3 wt% CB		5 wt% CB	
	MOR, MPa	MOE, GPa	**MOR, MPa**	MOE, GPa	**MOR, MPa**	MOE, GPa	**MOR, MPa**	MOE, GPa	**MOR, MPa**	MOE, GPa
ABS	**29-57**	1.5-3.1	**41(0.9)**	2.3(0.2)	**41(2)**	2.3(0.1)	**40(2)**	2.4(0.04)	**39(0.8)**	2.5(0.1)
HIPS	**16-30**	1.6-2.4	**33.7(0.8)**	2.2(0.04)	**33(2)**	2.17(0.1)	**33.5(1)**	2.2(0.03)	**32.6(0.9)**	2.2(0.03)
PC	**54-67**	1.9-2.7	**57.6(0.3)**	2.2(0.04)	**51(2)**	2.17(0.04)	**56(2)**	2.4(0.07)	**58.5(0.7)**	2.3(0.03)
PP	**16-31**	1-1.7	**32.7(1.7)**	1.8(0.1)	**32(2)**	1.9(0.04)	**32(0.6)**	1.9(0.06)	**32(1)**	2(0.06)

4. Conclusions

On the basis of the provided analysis, it can be concluded that characteristics of recycled material are comparable with related characteristics of their unprocessed counterparts. Elemental analysis showed the presence of additives, however, toxic chemicals were not detected. Thermal analysis showed the presence of some inorganic material. Melt flow indexes of recycled materials were comparable with virgin counterpart values found in the literature. In order to improve the mechanical properties of recycled material, carbon black (CB) filler was added. Results showed that CB has no significant effect on mechanical performance. Tensile samples fracture micrographs showed that CB particles were not uniformly dispersed in the polymer, mostly they stay aggregated.

References

[1] Baxter J, Wahlstrom, Castell-Rudenhausen MZ, Fråne A, Stare M, Lokke S and Pizzol M. Nordic Council of Ministers, 2014.
[2] Baxter J, Stensgår A, Raadal HL, Research O and Fråne A Nordic Council of Ministers, 2015 ANP 2015 p 718.
[3] Plastics-the Facts 2017. An analysis of European plastics production, demand and waste data. PlasticsEurope.
[4] Tange L, van Houwelingen JA, Hofland W and Salemis P In: Electronics Goes Green 2012+ (EGG), 2012 2012b. pp 1-6.
[5] Rothon RN Rapra Technology Ltd 2002 12(9) pp 13-15.
[6] Yousif BF, Low KO, Tan CM, Oi KH, Ong KL, Ng KL Proceedings of COMET 2006 14-15

IOP Conf. Series: Materials Science and Engineering **587** (2019) 012001 doi:10.1088/1757-899X/587/1/012001

January, UTM, Johor, Malaysia.
[7] Hayeemasae N, Ismail H, Khoon TB, Husseinsyah S and Harahap H 2016 Progress in Rubber, Plastic and Recycling Technology 32 pp 241-252.
[8] Lohar GS and Jogi BF 2018 *Procedia Manufacturing* 20 pp 85-90.
[9] Kanbur Y and Kucukyavuz Z 2009 *Journal of Reinforced Plastics and Composites* 20 2251-2260.
[10] Hall WJ and Williams PT *Journal of Analytical Applied Pyrolysis*, 79, 2007, pp 375-386.
[11] Troitzsch J 2004 Plastic Flammability Handbook. Principles, Regulations, Testing, and Approval 3rd Adition Carl Hanser Verlag, Munich, p 148.
[12] https://plastics.ulprospector.com.
[13] Jin H, Gonzales-Gutierrez J, Oblak P, Zupancic B and Emri I In Proceedings of the ANTEC, Cincinnati, OH, USA, 22-24 April 2013 pp 98-101.
[14] Puglisi C and Sturiale L 1999 *Macromolecules* 32 pp 2194-2203.
[15] Jang BM and Wilkie CA 2004 *Polymer Degradation Stability* 86 pp 419-430.

IOP Conf. Series: Materials Science and Engineering **587** (2019) 012002 doi:10.1088/1757-899X/587/1/012002

The biocompatibility of titanium alloy or stainless steel internal fixation material combined with anti-tuberculosis drug sustained-release carrier

Na Wang[1], Shaoning Guo[2], Kang Yang[3] and Jinguo Wang[4,*]

[1]Department of Anesthesiology, First Hospital of Jilin University, China
[2]Department of Postanesthesia Care Unit, First Hospital of Jilin University, China
[3]Department of Sports Medicine, First Hospital of Jilin University, China
[4]Department of Urology, First Hospital of Jilin University, China

*Corresponding author and e-mail: wangjinguolily@163.com

Abstract. The purpose of this paper is to summarize the treatment of characteristics of bone tuberculosis with titanium alloy or stainless steel internal fixation devices combined with anti-tuberculosis sustained-release materials. A search of literature on the implantation of titanium alloy or stainless steel internal fixation devices combined with anti-tuberculosis sustained-release materials for bone tuberculosis from 1995 to 2019 was performed using PubMed and Web of Science databases. Local sustained-release anti-tuberculosis drugs have been applied in surgical treatment of bone and joint tuberculosis. It can reduce the adverse reactions of systemic administration of a number of anti-tuberculosis drugs in patients with bone tuberculosis and continuously achieve the local effective anti-tuberculosis effect. At present, there are many kinds of materials used in drug sustained-release materials, each of which has its own advantages and disadvantages. It is difficult to achieve the ideal effect of anti-tuberculosis and bone repair. Different materials can be used together to make up the disadvantages for each other. The further progress towards ideal biological materials can be achieved. It may be the developing trend of clinical application of multi-material composite as sustained-release carrier against tuberculosis.

1. Introduction

With the development of internal fixation devices and anti-tuberculosis sustained-release materials, more and more scholars in China and at abroad begin to use titanium alloy and stainless steel internal fixation devices combined with anti-tuberculosis sustained-release materials to repair bone tuberculosis. Anti-tuberculosis drug sustained-release carriers have become a new highlight in the treatment of bone and joint tuberculosis in recent years [1]. After being implanted in the bone and joint tuberculosis lesions, the materials can be used as bone substitutes to repair bone defects and release anti-tuberculosis drugs for a long time at the stable concentration [2]. In the past, most clinical internal fixations were made of stainless steel, which is still widely used because of strength and price advantage [3]. Titanium is a safe metal element, very close to the elastic modulus of human bone, so it can produce good compatibility with human body mechanics after implantation. But titanium alloys are expensive, limiting its widespread use [4]. With the development of internal fixation devices and anti-tuberculosis drug sustained-release carriers, more and more domestic and foreign scholars begin

IOP Conf. Series: Materials Science and Engineering **587** (2019) 012002 doi:10.1088/1757-899X/587/1/012002

to use titanium alloy or stainless steel internal fixation devices combined with anti-tuberculosis drug sustained-release carriers to repair spinal tuberculosis [5]. In this paper, the characteristics of repairing bone tuberculosis lesions with titanium alloy or stainless steel internal fixation devices combined with anti-tuberculosis drug sustained-release carriers are reviewed.

2. Methods

2.1. Source of data
Articles on titanium alloy or stainless steel internal fixation devices combined with anti-tuberculosis drug sustained-release carriers for bone tuberculosis lesion from January 1995 to February 2019 on PubMed database were indexed by computer for "bone tuberculosis, titanium alloy or stainless steel, fixation, released anti-tuberculosis materials".

2.2. Inclusion and exclusion criteria

2.2.1. Inclusion criteria
Research articles related to the research contents of sustained release drugs against bone tuberculosis. The contents of the literature are representative and the data are reliable.

2.2.2. Exclusion criteria
Poor relevance with the purpose of the study and old, repetitive literature.

3. Results
A total of 129 papers were retrieved, among which 26 literatures with poor correlation with the research purpose and obsolete and repeated contents were excluded, and 103 papers meeting the standards were included for review.

4. Discussion

4.1. Titanium alloy or stainless steel internal fixation materials
In the past, most clinical internal fixations were made of stainless steel. The stainless steel material has good strength, but its elastic modulus is too large to be well matched with the elastic modulus between bones, which may lead to various adverse consequences. Titanium is a safe metal element, very close to the elastic modulus of human bone, so it can produce good compatibility with human body mechanics after implantation. Moreover, titanium alloy has strong inertia and corrosion resistance, not easy to appear a variety of infections. Therefore, titanium alloy is widely used in clinical treatment of various diseases, and play a good role. The experiment compared the effect and biocompatibility of titanium alloy and stainless steel internal fixation material in the treatment of spinal tuberculosis. But stainless steel internal fixation materials due to its price factor is also widely used [6].

4.2. Sustained release carriers of artificial bone drugs
Artificial bone is mainly calcium and phosphorus materials, including hydroxyapatite, tricalcium phosphate, bioglass, calcium phosphate bone cement, calcium sulfate, etc. The chemical composition of such inorganic compounds is similar to that of human bones and can be used as drug sustained-release carriers. In vivo degradable, non-toxic, with good mechanical strength and bone conduction performance. The material has porous structure and can support multiple drugs, bone growth factors and so on. Drug release rate can be controlled by adjusting the pore size of the material itself and the content of additives. The disadvantage of using artificial bone alone is that the PH value of local microenvironment is affected after degradation. The plasticity of ceramic artificial bone is poor, and the degradation speed of artificial bone material is slow, which is not matched with the speed of bone repair. Some scholars developed an implantable sustained-release agent of rifampin-porous hydroxyapatite artificial bone nucleus for the treatment of bone tuberculosis, which was in the shape

IOP Conf. Series: Materials Science and Engineering **587** (2019) 012002 doi:10.1088/1757-899X/587/1/012002

of a hollow cylinder [7]. The experiment confirmed that the drug with an effective concentration could be continuously released in the body for at least 27 weeks, and it had an obvious bacteriostatic effect on mycobacterium tuberculosis. Medical calcium sulfate granules containing tobramycin or a mixture of autologous bone and calcium sulfate granules were used for the repair of bone defects after the surgery of bone and joint tuberculosis [8].

4.3. Sustained release materials of polylactones

With polylactic acid (PLA), poly (lactic acid - glycolic acid (PLGA) as the representative of aliphatic polyester synthetic polymer because by chemical synthesis, production, product performance can control, repeatability is good, suitable for the requirement of engineering, but also by changing the structure of the synthesis of polymer, composition to control mechanical properties and biodegradation rate, to meet the needs of different biomedical applications [9]. Them inside the organism and the ester bond rupture due to hydrolysis reaction, degradation, degradation products of lactic acid and glycolic acid but also through the Krebs cycle further isn't toxic metabolic tissue in pairs of carbon dioxide and water, and thus have obtained approval in China and the United States and other countries certification, can be used for the preparation of implanted medical material. After implantation, the material has certain initial mechanical properties. The degradation rate of mechanical properties of degradable materials matched with the rate of bone healing [10]. A series of studies were conducted on the cylindrical PLGA material loaded with isoniazid to study the characteristics of drug release and the repair effect on bone defects. It was indicated that the slow release material system had the ability of local slow release of isoniazid for more than 6 weeks, and the slow release and degradation rate of the material matched the speed of bone repair. However, polylides lack good hydrophilicity, biological activity and sufficient mechanical strength, and generate a large number of low molecular products with carboxyl group after hydrolysis, which can reduce the PH value of the biological internal environment. Low PH value will cause inflammation and damage to the body [11].

4.4. Natural bone materials

The natural bone materials used in clinical practice are mainly produced by the processing of allogeneic bone, including freeze-dried bone, calcified bone matrix, deproteinated bone matrix, acellular bone matrix, etc., with bone conduction and certain bone induction activity [12]. With porous structure and regular pore size, it can load many kinds of drugs. However, as a drug carrier alone, there are still some shortcomings, such as insufficient bone strength, residual immunogenicity, small drug load and poor plasticity [13]. The concentration of anti-tuberculosis drugs plays a vital role in killing mycobacterium tuberculosis. Anti-tuberculosis sustained-release drugs can significantly improve the local drug concentration and enhance the effect of local anti-tuberculosis drugs. Pharmacokinetic studies have proved that maintaining the sustained and stable effective bactericidal concentration around the lesion is the key to killing tuberculosis bacillus. Blood drug concentrations of INH, RFP and PZA almost reach or close to the peak at about 2h after taking the drug, and enter the drug clearance phase at 4-6h. Traditional drug administration makes it difficult for anti-tuberculosis drugs to play a sustained bactericidal role in the local lesion, which is one of the reasons for poor efficacy of spinal tuberculosis and easy to produce drug resistance [14]. Maintaining the local concentration of anti-tuberculosis drugs in the lesion greater than the minimum bactericidal concentration is the key to the treatment of spinal tuberculosis.

4.5. Carriers of multiple materials

Composite material system with single carrier material as joint tuberculosis slow-release drug system used to repair bone defect, and there may be a poor mechanical strength, bone conduction ability only, do not have ability of bone induction, poor plasticity, poor hydrophilicity, microstructure affects cell adhesion, the influence microenvironment PH value, etc, to reach the ideal treatment of joint tuberculosis have ability to repair bone defect [15]. In recent years, the research and application of composite material systems that combine two or more bioactive materials and concentrate the advantages of different materials into one material have attracted extensive attention from scholars at home and abroad. Therefore, sustained-release materials containing anti-tuberculosis drugs are introduced into the treatment of spinal tuberculosis. Makino et al. embedded the RFP package in the microsphere and found that its release cycle could reach 20 days. Wu et al. developed drug-loaded artificial bone that can carry multiple drugs and has the characteristics of controlled release [16]. The controlled release drug-loaded artificial bone has the characteristics of controlled release and sustained release of drugs. It can achieve the treatment of combined drug therapy and bone defect repair at the same time. However, these studies have shortcomings of single drug delivery or failure to describe in vitro anti-tb performance. Currently, materials for bone defect reconstruction after spinal tuberculosis surgery mainly include autologous bone, allogeneic bone, bone repair materials, etc. [17], among which, autologous bone and allogeneic bone have achieved good results in clinical application [18], but their application is limited due to limited sources and the possible spread of infectious diseases. Bone repair materials shorten the time of bone formation and improve the quality of bone formation, and bone repair materials as the carrier of drug sustained-release system can achieve the dual purpose of bone defect repair and drug treatment.

5. Conclusions

The ideal local sustained-release drug-carrier system for bone and joint tuberculosis should have good biocompatibility, good affinity for cells and high adhesion rate of cells to the material. The materials have bone conductivity and bone inductivity in vivo. It has biodegradable property, high biodegradable matching rate with bone without immunogenicity. The material and its degradation products have no adverse reaction to the body. It has good plasticity, suitable size, porosity and mechanical strength, good physical and chemical properties. It is convenient for medical disinfection. It has the function of slow-release drugs, and there is no adverse interaction between carriers and drugs. The degradation time of sustained-release carriers is long, and the drug effect is maintained for more than 2-3 months. Multiple first-line anti- tuberculosis drugs can be administered together with the carriers, and the concentration of local sustained-release drugs was stable.

Acknowledgement

This research was financially supported by the First Hospital of Jilin University.

References

[1] Bohner M. New hydraulic cements based on alpha-tricalcium phosphate-calcium sulfate dihydrate mixtures. Biomaterials. 2004; 25: 741-749.

[2] Fei Z, Hu Y, Wu D, et al. Preparation and property of a novel bone graft composite consisting of rhBMP-2 loaded PLGA microspheres and calcium phosphate cement. J Mater Sci Med. 2008; 19(3): 1109-1116.

[3] Bigi A, Bracci B, Panzavolta S. Effect of added gelatin on the properties of calcium phosphate cement. Biomaterials. 2004; 25: 2893-2899.

[4] Chen G, Ushida T, Tateishi T. A biodegradable hybrid sponge nested with collagen microsponges. J Biomed Mater Res. 2000; 51: 273-279.

[5] Jaganathan KS, Rao YU, Singh P, et al. Development of a single dose tetanus toxoid formulation based on polymeric microspheres: a comparative study of poly(D, L-lactic-co-glycolic acid) versus chitosan microspheres. Int J Pharm. 2005; 294: 23-32.

[6] Dutt M, Khuller GK. Chemotherapy of Mycobacterium tuberculosis infections in mice with a combination of isoniazid and rifampicin entrapped in Poly (DL-lactide-co-glycolide) microparticles. J Antimicrob Chemother. 2001; 47(6): 829-835.

[7] Fahmy TM, Samstein RM, Harness CC, et al. Surface modification of biodegradable polyesters with fatty acid conjugates for improved drug targeting. Biomaterials. 2005; 26: 5727-5736.

[8] Yao J, Radin S, SLeboy P, et al. The effect of bioactive glass content on synthesis and

bioactivity of composite poly (lactic-co-glycolic acid)/bioactive glass substrate for tissue engineering. Biomaterials. 2005; 26: 1935-1943.

[9] Xu HH, Quinn JB, Takagi S, et al. Synergistic reinforcement of in situ hardening calcium phosphate composite scaffold for bone tissue engineering. Biomaterials. 2004; 25: 1029-1037.

[10] Simon CG Jr, Khatri CA, Wight SA, et al. Preliminary report on the biocompatibility of a moldable, resorbable, composite bone graft consisting of calcium phosphate cement and poly(lactide-co-glycolide) microspheres. J Orthop Res. 2002; 20: 473-482.

[11] Gao Y, Zuo J, Bou-Chacra N, et al. In vitro release kinetics of antituberculosis drugs from nanoparticles assessed using a modified dissolution apparatus. Biomed Res Int. 2013; 2013: 136590.

[12] Gelperina S, Kisich K, Iseman MD, et al. The potential advantages of nanoparticle drug delivery systems in chemotherapy of tuberculosis. Am J Respir Crit Care Med. 2005; 172(12): 1487-1490.

[13] Yang XB, Roach HI, Clarke NM, et al. Human osteoprogenitor growth and differentiation on synthetic biodegradable structures after surface modification. Bone. 2001; 29: 523-531.

[14] du Toit LC, Pillay V, Danckwerts MP. Tuberculosis chemotherapy: current drug delivery approaches. Respir Res. 2006; 7: 118-122.

[15] Hussain A, Singh S, Das SS, Anjireddy K, Karpagam S, Shakeel F. Nanomedicines as Drug Delivery Carriers of Anti-Tubercular Drugs: From Pathogenesis to Infection Control. Curr Drug Deliv. 2019 Feb 1. doi: 10.2174/1567201816666190201144815.

[16] Singh G, Kesharwani P, Srivastava AK. Tuberculosis Treated by Multiple Drugs: An Overview. Curr Drug Deliv. 2018; 15(3): 312-320.

[17] Athar MT, Ullah Z. Current Approaches and Future Prospects of Nanomedicine in Tuberculosis Therapy. Recent Pat Antiinfect Drug Discov. 2017; 12(2): 120-127.

[18] Walker KB, Guo M, Guo Y, Poecheim J, Velmurugan K, Schrager LK. Novel approaches to preclinical research and TB vaccine development. Tuberculosis (Edinb). 2016; 99 Suppl 1: S12-15.

IOP Conf. Series: Materials Science and Engineering **587** (2019) 012003 doi:10.1088/1757-899X/587/1/012003

The sustained-release effect of triple anti-tuberculosis drugs carried by calcium sulfate/poly-amino acid compound materials in bone tuberculosis lesion

Na Wang[1], Yongyan Zou[2], Kang Yang[3] and Jinguo Wang [4*]

[1]Department of Anesthesiology, First Hospital of Jilin University, China
[2]Department of Anesthesiology, Changchun Center Hospital, China
[3]Department of Sports Medicine, First Hospital of Jilin University, China
[4]Department of Urology, First Hospital of Jilin University, China

*Corresponding author and e-mail: wangjinguolily@163.com

Abstract. This article is to review the research progress of the sustained-release effect of triple anti-tuberculosis drugs carried by calcium sulfate/poly-amino acid compound in bone tuberculosis lesion, and to investigate the biocompatibility, sustained-release and anti-tuberculosis properties of calcium sulfate/poly-amino acid compound materials from the aspects of tissue engineering modification. Papers from January 2000 to February 2019 are retrieved in PubMed database and Web of Science database. Searching words are "bone tuberculosis, biocompatibility, sustained release, anti-tuberculosis, calcium sulfate/poly-amino acid". A total number of 26 articles were included. After the artificial sustained release system of calcium sulfate/poly-amino acid containing triple anti-tuberculosis drugs was put into the bone tuberculosis lesion, the anti-tuberculosis drugs could be released slowly and continuously for up to three months according to results of the related papers. This locally sustained release effect is caused by the direct diffusion of anti-tuberculosis drugs from the carriers and the dispersion of drugs caused by the exposure of drug groups between calcium sulfate crystals after the degradation of the artificial material. The drug concentrations of the three anti-tuberculosis drugs in the bone and lumbago major muscle tissues at each detection time point were higher than those in venous blood. Three anti-tuberculosis drugs carried by calcium sulfate/poly-amino acid compound can be simultaneously released and sustained-released into the bone lesion. It is more efficient and faster to kill tuberculosis bacillus than the single-drug slow-release material which was previously studied, and consistent with the principle of using anti-tuberculosis drugs.

1. Introduction

In view of the barrier effect of the lesion wall, it is difficult to reach the effective drug concentration in the local lesion of spinal tuberculosis through oral and intravenous administration, which often leads to the uncured or recurrent treatment of spinal tuberculosis. How to improve the concentration of local anti-tuberculosis drugs in the lesion is of great significance for improving the efficacy of spinal tuberculosis [1]. Drug release system implantation is the focus of current spinal tuberculosis research, and many studies have confirmed that calcium sulfate/poly-amino acid can be used as bone repair materials. The material after implantation, except with filling and osteogenesis effect for composite

IOP Conf. Series: Materials Science and Engineering **587** (2019) 012003 doi:10.1088/1757-899X/587/1/012003

degradation process in absorption phase is different, the artificial materials so the load of drugs can be sustained, effective slow release, in order to improve the lesion partial effective anti-tb drug concentration, reduce drug adverse reaction, reduce the emergence of resistant strains, make further shortened the chemotherapy treatment, improve curative effect [2]. The physical and chemical properties, in vitro and in vivo sustained release properties and other aspects of the artificial sustained release materials of calcium sulfate/poly-amino acid containing triple antituberculosis drugs have been deeply studied, and its good local sustained release properties have been verified [3].

2. Methods

2.1. Source of data

Articles on the sustained-release effect of triple anti-tuberculosis drugs carried by calcium sulfate/poly-amino acid compound in bone tuberculosis lesion from January 2000 to February 2019 on PubMed and Web of Science database were indexed by computer for "bone tuberculosis, biocompatibility, sustained release, anti-tuberculosis, calcium sulfate/poly-amino acid".

2.2. Inclusion and exclusion criteria

2.2.1. Inclusion criteria

The sustained-release materials of calcium sulphate-polyamino acid containing triple anti-tuberculosis drugs and the study on treatment of bone tuberculosis lesion were retained and summarized.

2.2.2. Exclusion criteria

The articles with poor correlation with the research objective and outdate and repeated contents were excluded.

3. Results

A total of 39 papers were retrieved, among which 13 literatures with poor correlation with the research purpose and obsolete and repeated contents were excluded, and 26 papers meeting the standards were included for review. After the artificial sustained release system of calcium sulfate/poly-amino acid containing triple anti-tuberculosis drugs was put into the bone tuberculosis lesion, the anti-tuberculosis drugs could be released slowly and continuously for up to three months according to results of the related papers. This locally sustained release effect is caused by the direct diffusion of anti-tuberculosis drugs from the carriers and the dispersion of drugs caused by the exposure of drug groups between calcium sulfate crystals after the degradation of the artificial material. The drug concentrations of the three anti-tuberculosis drugs in the bone and lumbago major muscle tissues at each detection time point were higher than those in venous blood.

4. Discussion

4.1. Preparation and implantation of the sustained-release material

According to the literature, calcium sulfate, amino acids, isoniazid, pyrazinamide and rifampicin in 9:1:0.3:0.3:1.2 the proportion of mixed, preparation of carrying the joint the chemotherapy medicine calcium sulphate/poly (amino acid material of artificial bone, are preparing to grow 5 mm, 3.5 mm wide, 3.5 mm high rule of bone, using electronic balance measured containing triple the chemotherapy medicine calcium sulphate/poly (amino acid artificial slow-release materials of bone quality is 144.24 mg, pyrazinamide contained, isoniazid and rifampicin respectively is 14.67, the quality of 3.67, 3.67 mg. Artificial bone materials without calcium sulfate/poly-amino acid were prepared by the same method, and no anti-tuberculosis drugs were added. The implantation of sustained-release materials: after the tuberculosis lesion was cleared, the bone groove was implanted with the artificial bone sustained-release materials containing calcium sulfate/poly-amino acid, a triple anti-tuberculosis drug [4-6].

4.2. Selection of anti-tuberculosis drugs in the sustained-release implantation system

Clinical anti-tuberculosis drugs include streptomycin, rifampicin, isoniazid, ethylamine butanol, pyrazinamide, etc., and streptomycin is the most commonly used drug for the traditional focus after the elimination of bone and joint tuberculosis. Although streptomycin is broad-spectrum antibiotic, but in vivo it is difficult to achieve bactericidal effect, mainly play a role in bacteriostasis. It is not easy to enter the cell membrane and cheese tissue, poor effect on bone tuberculosis, fibrous cavity type tuberculosis [7]. And because of the serious consequence of streptomycin allergy, it has certain limitation in clinical use. Pyrazinamide is an effective oral anti-tuberculosis drug, which has a special killing effect on the mycobacterium tuberculosis in the cell or in the static state. It has enhanced antibacterial effect in the acidic environment, but it is more toxic. At present, rifampicin and isoniazid are often selected in the study of sustained release of anti-tuberculosis drugs. The reproduction of bacteria has bactericidal effect [8]. To the tuberculosis bacilli of quiescent period, raise medicaments concentration or prolonging contact time also can have antiseptic effect, and have coequal to the tuberculosis bacilli inside and outside the cell kill [9].

4.3. Requirement of the drugs and carriers for bone tuberculosis lesion

Tuberculosis focal tissue of bone and joint includes caseous necrotic material, dead bone, cold abscess, tuberculosis granulation tissue, and mycobacterium tuberculosis flora. In non-sclerosing lesions, the periphery is surrounded by fibrous connective tissue. In sclerosing lesions (approximately 70% of spinal tuberculosis), the periphery of the lesion is a sclerotic wall, namely sclerotic bone [10]. In non-sclerosing lesions, the concentration of first-line anti-tb drugs can reach above the minimum inhibitory concentration. However, in the sclerosing type, the sclerosed bone limited the development of the lesion, but there was almost no blood supply [11]. The anti-tuberculosis drugs in the sclerosed bone were only at the minimum inhibitory concentration, which could not reach the bactericidal concentration, and formed a barrier to prevent the drugs from entering the lesion. However, the bacteria hiding in necrosis and caseous lesions were only sensitive to rifampicin and pyrazinamide. Extremely low drug concentration, relatively closed environment of the lesion, hypoxia and other complex factors make it difficult to kill the bacteria in sclerosed osteonecrosis and caseous lesions, which become potential factors for tuberculosis recurrence [12]. Therefore, the drug is directly placed in the lesion through local drug delivery, and anti-tuberculosis drugs are locally released, which is conducive to the treatment of tuberculosis. The advantage of the local drug sustained-release system for bone and joint tuberculosis lies in that it can achieve a higher local bactericidal concentration, which is conducive to killing or inhibiting the growth and reproduction of mycobacterium tuberculosis and reducing drug resistance caused by low local drug concentration. Blood drug concentration is low, systemic adverse reaction is small [13]. To avoid the scar tissue and bone sclerosis caused by surgery to prevent the anti-tuberculosis drugs from reaching the focus. After the implantation of the lesion, the drug carrier material will be gradually degraded by the body and the drug will be released and eventually replaced by bone tissue [14].

After the artificial sustained release material of calcium sulfate/poly-amino acid containing triple anti-tuberculosis drugs was put into the bone lesion defect of the spinal tuberculosis model, the anti-tuberculosis drugs could be released slowly and continuously for up to three months. Locally released drugs not only enter the bone and the lumbar maximus tissue on its surface, but also part of them can be absorbed into the blood circulation of the whole body [15]. This locally sustained release effect is caused by the direct diffusion of anti-tuberculosis drugs from the surface of the artificial material and the dispersion of drugs caused by the exposure of drug groups between calcium sulfate crystals after the degradation of the artificial material. The drug concentrations of the three anti-tuberculosis drugs in the bone and lumbago major muscle tissues at each detection time point were higher than those in venous blood. Three anti-tuberculosis drugs can simultaneously be released from the slow-release material and enter the local lesion site. Compared with the single-drug slow-release material previously studied, it can kill tuberculosis bacillus more efficiently and rapidly, which is more consistent with the principle of using anti-tuberculosis drugs.

5. Conclusions

It is believed that calcium sulfate/poly-amino acid sustained-release materials containing triple anti-tuberculosis drugs can be directly implanted into the bone tuberculosis lesions, which can effectively improve the concentration of local anti-tuberculosis drugs, significantly improve the antibactericidal efficiency, shorten the treatment cycle, significantly improve the curative effect, and enable patients to recover at an early date. However, the material still has problems such as insufficient compressive strength and rapid drug release. With the effective solution of these problems, the artificial sustained-release materials will have a broader application prospect in clinical practice.

Acknowledgement

This research was financially supported by the First Hospital of Jilin University.

References

[1] Ruhé PQ, Boerman OC, Russel FG, Spauwen PH, Mikos AG, Jansen JA. Controlled release of rhBMP-2 loaded poly(dl-lactic-co-glycolic acid)/calcium phosphate cement composites in vivo. J Control Release. 2005; 106(1-2): 162-171.

[2] Ruhe PQ, Hedberg EL, Padron NT, Spauwen PH, Jansen JA, Mikos AG. rhBMP-2 release from injectable poly(DL-lactic-co-glycolic acid)/calcium-phosphate cement composites. J Bone Joint Surg Am. 2003; 85-A Suppl 3: 75-81.

[3] Li M, Liu XD, Liu XY, Ge BF. Preparation and ectopic osteoinduction study of macroporous bone substitute with calcium phosphate cements and rhBMP-2 loaded gelatin microspheres. Zhongguo Gu Shang. 2011; 24(5): 411-415. Chinese.

[4] Masaeli R, Jafarzadeh Kashi TS, Dinarvand R, Rakhshan V, Shahoon H, Hooshmand B, Mashhadi Abbas F, Raz M, Rajabnejad A, Eslami H, Khoshroo K, Tahriri M, Tayebi L. Efficacy of the biomaterials 3wt%-nanostrontium-hydroxyapatite-enhanced calcium phosphate cement (nanoSr-CPC) and nanoSr-CPC-incorporated simvastatin-loaded poly(lactic-co-glycolic-acid) microspheres in osteogenesis improvement: An explorative multi-phase experimental in vitro/vivo study. Mater Sci Eng C Mater Biol Appl. 2016; 69: 171-183.

[5] Blom EJ, Klein-Nulend J, Wolke JG, van Waas MA, Driessens FC, Burger EH. Transforming growth factor-beta1 incorporation in a calcium phosphate bone cement: material properties and release characteristics. J Biomed Mater Res. 2002; 59(2): 265-272.

[6] Liu HW, Chen CH, Tsai CL, Hsiue GH. Targeted delivery system for juxtacrine signaling growth factor based on rhBMP-2-mediated carrier-protein conjugation. Bone. 2006; 39(4): 825-836.

[7] Schrier JA, DeLuca PP. Recombinant human bone morphogenetic protein-2 binding and incorporation in PLGA microsphere delivery systems. Pharm Dev Technol. 1999; 4(4): 611-621.

[8] Seeherman H. The influence of delivery vehicles and their properties on the repair of segmental defects and fractures with osteogenic factors. J Bone Joint Surg Am. 2001; 83 (suppl1): 61-79

[9] Sia IG, Wieland ML. Current Concepts in the Management of Tuberculosis. Mayo Clin Proc. 2011; 86(4): 348-361.

[10] Fei Z, Hu Y, Wu D, Wu H, Lu R, Bai J, Song H. Preparation and property of a novel bone graft composite consisting of rhBMP-2 loaded PLGA microspheres and calcium phosphate cement. J Mater Sci Mater Med. 2008; 19(3):1109-1116.

[11] Ruckh TT, Oldinski RA, Carroll DA. Antimicrobial effects of nanofiber poly(caprolactone)

tissue scaffolds releasing rifampicin. J Mater Sci Mater Med. 2012; 23(6): 1411-1420.
[12] Wang Z, Ge Z, Jin W, et al. Treatment of spinal tuberculosis with ultrashort- course chemotherapy in conjunction with partial excision of pathologic vertebrae.Spine J. 2007; 7(6): 671-681.
[13] Diacon AH, Donald PR. The early bactericidal activity of antituberculosis drugs. Expert Rev Anti Infect Ther. 2014; 12(2): 223-237.
[14] Vora C, Patadia R, Mittal K, et al. Recent patents and advances on anti-tuberculosis drug delivery and formulations. Recent Pat Drug Deliv Formul. 2013; 7(2): 138-149.
[15] Turner TM, Urban RM, Gitelis S, et al. Resorption evaluation of a large bolus of calcium sulfate in a canine medullary defect. Orthopedics. 2003; 26(5 Suppl): s577-579.

IOP Conf. Series: Materials Science and Engineering **587** (2019) 012004 doi:10.1088/1757-899X/587/1/012004

Development and application of an anticorrosion type water release agent

Juxiang Xing[1*], Long Xiong[2], Xing Li[3]

Building 12th, China University of Geosciences, Science Park, Wuhan, China

*Corresponding author and e-mail: Juxiang Xing, xingjuxiang@126.com

Abstract. Water based release agents have attracted widely attention in recent years due to their advantages, such as environmental and convenience. However, during the application process, the resistance to the steel mould is not obvious, and the concrete surface is easy to rust, which affects the appearance of the concrete surface. The application of aluminium mould can't solve the problem of bubble and surface ash. The ZJTM anticorrosion type water release agent developed in this paper can effectively solve the corrosion of the concrete surface, bubble and surface ash, and the effect is remarkable in all kinds of engineering.

1. Introduction

Concrete release agent, also known as concrete isolator or release lubricant. It is a kind of material that can be coated or sprayed on the inner surface of formwork to produce a layer of isolation film, which can effectively reduce the adhesion between concrete and formwork, and promote the smooth removal of concrete from formwork and keep the shape of concrete.

As a supporting material for concrete construction, Concrete release agent has almost the same history with concrete. In the nineteen-twenties, soap water solution, waste diesel oil, waste oil, lime, talc powder were used as releasing agent in concrete engineering in various countries, but the effect was general, and soap water solution was alkaline solution, which affected the surface quality of concrete. With the application of new concrete pouring technology, the emergence of new formwork such as steel formwork, FRP formwork, wood formwork, aluminium alloy formwork, reinforced concrete formwork and plastic formwork, and the improvement of concrete surface quality and decoration grade, new and higher requirements for Concrete release agent are put forward [1]. Since the early 1960s, after the advent of foreign commercial concrete release agents, great progress has been made in the research of concrete release agent for concrete moulding. The development of concrete release agent is very rapid, and a series of products have been formed, which are gradually matched with modern construction and template technology.

In recent years, water-borne release agent has attracted wide attention because of its environmental protection, easy to pollute the environment, easy to use and other advantages [2], However, the embroidery resistance of steel formwork is not obvious in the application process, and the concrete surface is easy to rust, which affects the beauty of the concrete surface; the application of aluminum oxide template can't solve the problem of air bubbles and surface ash.

According to the application problem of water-borne release agent, we developed a new kind of emulsifying release agent product: ZJTM water-borne release agent for corrosion inhibition of concrete. The test shows that the product with inositol hexaphosphate is stable, environmentally friendly, harmless to the body, film-forming effect is good, easy to demoulding, and can effectively

IOP Conf. Series: Materials Science and Engineering **587** (2019) 012004 doi:10.1088/1757-899X/587/1/012004

solve the corrosion and bubble on the surface of concrete after demoulding of various metal formwork. As well as the problem of ashing, the application effect is remarkable in all kinds of projects.

2. Experimentation

2.1. Major raw materials and instruments
Film forming agent, sorbitol trioleate, diethylene glycol monostearate, polyoxyethylene sorbitol monolaurate, 1,3-propylene glycol, triethanolamine, sorbitol monooleate, thiophosphate, inositol hexaphosphate, imidazoline oleate, water-borne concrete release agent 1, water-borne concrete release agent 2.

Fasks (1L), electric agitator (PTFE agitator), centrifuge, electronic balance, vacuum dryer, beaker.

2.2. Preparation of Rust-inhibiting Demoulding Agent
The rust-inhibiting water-borne release agent is composed of film-forming agent, emulsifier, dispersant, water and corrosion inhibitor. In this method, the selection and proportion of corrosion inhibitor is the key. The specific experimental steps are: (1) adding film-forming agent to four flasks at one time, stirring and heating to 60 degrees; (2) adding compound emulsifier with HLB value of 9-10 at one time, stirring for 10 minutes at 60°C; (3) adding 65 degrees hot water to phase transition continuously and slowly, stirring for 30 minutes after adding; (4) adding compound emulsifier at last. Different corrosion inhibitors were stirred for 10 minutes to obtain water-borne rust-inhibiting release agent (effective ingredient content 12% [3].

2.3. Performance testing

2.3.1. Material and Ratio for Performance Testing
a. Cement

According to the use of cement in concrete production enterprises, Huaxin P.O 42.5 cement and Yadong P.O 42.5 cement are selected. The basic properties are shown in Table 1.

Table 1. Basic properties of test cement

Sample	Water consumption for standard consistency /g	Initial setting/min	Final setting /min	Density /(g/cm^3)	Specific surface area /(cm^2/g)	3d mortar strength /MPa	28d mortar strength /MPa
Huaxin Cement	124	155	220	3.08	4055	30.0	50.1
Ya Dong cement	132	185	236	3.02	3380	26.5	51.9

b. Mineral powder

S95 mineral powder (KF) from a new building material plant has a specific surface area of 4250 cm^2/g, a density of 2.86 g/cm^3, an activity of 86% on 7d and 97% on 28d.

c. Fly ash

The grade II fly ash is used with fineness of 2.5%, loss of combustion of 8.0% and water requirement ratio of 105%.

d. Sand

Sand: its fineness modulus is 2.6, the mud content is 0.8%, and the mud content is 0.3%.

e. Gravel

C30 concrete adopts 5-31.5 mm continuous Graded Macadam with apparent density of 2730 kg/m^3, mud content of 0.4%, mud content of 0.1%, needle-like content of 4.0%, crushing index of 6.5%.

IOP Conf. Series: Materials Science and Engineering **587** (2019) 012004 doi:10.1088/1757-899X/587/1/012004

C70 concrete adopts 5-25 mm continuous Graded Macadam with apparent density of 2730 kg/m^3, mud content of 0.4%, mud content of 0.1%, needle-like content of 4.0%, crushing index of 6.5%.

f. Water

Tap water or groundwater meet the requirements of JGJ 2006 Water Standard for Concrete Mixing.

h. Admixture

The admixture is the finished product with 10% solid content produced by the company, which meets the requirements of GB 8076-2008.

Table 2 The concrete mix

Strength grade	Cement	Fly ash	Mineral powder	Sand	Gravel	Water
C30	190	100	85	860	1020	135
C70	450	50	115	700	960	120

2.3.2. Performance testing methods

- Emulsion stability test: according to JC/T 949-2005 "release agent for concrete products" for testing. The product was diluted to 25 mL of the diluent of the concentration used and injected into the colorimetric tube. The product was stationary for 24 hours at 5-40°C. The homogeneity and stratification of the product under natural light were observed.
- Storage stability test: put the emulsion into the 10 mL centrifuge tube and put it into the high-speed centrifuge, set the speed of 4000r/min, centrifuge the 10min, and evaluate according to the 6 grades of emulsion stability.
- Demoulding performance test: refer to Appendix A of JC/T 949-2005 "Demoulding Agents for Concrete Products".

3. Results and Discussions

3.1. Rust Inhibiting Performance Test of Different Corrosion Inhibitors

The application of water-borne release agent in steel moulds for construction generally has some problems, such as easy corrosion on the surface of steel moulds and easy contamination of concrete surface by rust. According to the mechanism of electrochemical corrosion of steel moulds and the effect of film formation on the surface of steel templates, triethanolamine, sorbitol monooleate, thiophosphate, inositol hexaphosphate, oleic acid imidazoline and other substances were selected to carry out orthogonal experiments from Table 3. It can be seen that inositol hexaphosphate and imidazoline oleate have good antirust effect. The inositol hexaphosphate molecule has 24 oxygen atoms, 12 hydroxyl groups and 6 phosphate groups which can cooperate with metal. It is a rare metal polydentate mixture. Only one of the six phosphate groups in the molecular structure is in a position, and the other five are in E position. Four phosphoric groups are on the same plane. Therefore, when inositol hexaphosphate complexes with metals on the metal surface, it is easy to form a dense single molecule protective film on the metal surface, which can effectively prevent 02 from entering the metal surface, thus slowing down the corrosion of metals [4].

Imidazoline corrosion inhibitors are widely used in various anti-corrosion fields because of their advantages of green, environmental protection and low toxicity. Compared with other types of organic inhibitors, imidazoline inhibitors can complex some oxidants in solution to reduce the electrode potential of metals to achieve the purpose of corrosion inhibition [5].

IOP Conf. Series: Materials Science and Engineering **587** (2019) 012004 doi:10.1088/1757-899X/587/1/012004

Table 3. Test results of different rust inhibitors

Sample	Emulsion stability	Storage period /d	Film forming time /min	Erosion resistance	Corrosion resistance (Rust spots appear)
-	UN stratified	200	10	no	2d
triethanolamine	UN stratified	150	16	no	2d
sorbitan monooleate	UN stratified	200	15	no	5d
Thiophosphate	UN stratified	120	14	no	2d
Inositol hexaphosphate	UN stratified	210	18	yes	30d
Oleic imidazoline	UN stratified	180	12min	yes	30d
Water-borne release agent 1	UN stratified	90	20min	no	1d
Water-borne release agent 2	Slight bleaching oil	30	18min	no	1d

3.2. Effect of Different Corrosion Inhibitors on the Appearance of Concrete after Demoulding

In the construction of concrete engineering, honeycomb pitting surface, lack of angle, bars and other quality defects often appear on the surface of concrete components, which not only affects the apparent effect of concrete, but also affects the internal quality and durability of concrete seriously. With the development of new technology and technology of concrete, the requirement for decorative effect of concrete surface is higher and higher. The use of demoulding agent can reduce or even avoid the above problems on the surface of concrete components, and improve the quality of the surface of components.

Combining with the basic performance test of demoulding agent, three groups of better performance anti-rust demoulding agent were selected to compare with oil demoulding agent. Fig. 1 is the application of different demoulding agent in experimental pier column. From the figure, it can be seen that the concrete using water-based demoulding agent is obviously superior to oil demoulding agent. At the same time, inositol hexaphosphate rust-inhibiting demoulding agent is coagulated compared with other anti-rust demoulding agents. There is no rust on the surface of the soil, no harmful stomata and uniform color.

IOP Conf. Series: Materials Science and Engineering **587** (2019) 012004 doi:10.1088/1757-899X/587/1/012004

a. Oily release agent

b. Oleic acid imidazoline rust-inhibiting release agent

c. Sorbitol monooleate rust-inhibiting release agent

d. Inositol hexaphosphate rust-inhibiting release agent

Figure 1. Application of different demoulding agents in experimental piers

3.3. Adaptability of Different Inhibitors to Template

At present, formwork in construction industry mainly includes wood formwork, steel formwork, plastic formwork, aluminum alloy formwork, etc. Among them, aluminum alloy formwork is more and more widely used because of its light weight, convenient installation and disassembly, high turnover times, wide application environment, good durability, less garbage in construction site and high recycling value. However, the aluminum alloy formwork is assembled as a whole, and the formwork is easy to react with concrete, the wall surface is easy to peel off after demoulding, bubbles and other issues. Figure 2 compares the appearance of concrete after different rust-inhibiting demoulding agents are applied to the aluminum alloy formwork. It can be seen that inositol hexaphosphate rust-inhibiting demoulding agent has significantly improved the ash and bubbles on the concrete surface, which benefits from this. The corrosion inhibitor can form a dense protective film on the surface of the formwork to prevent the reaction between the aluminium alloy formwork and fresh concrete.

a. Oily release agent

b. Oleic acid imidazoline rust-inhibiting release agent

c. Sorbitol monooleate rust-inhibiting release agent

d. Inositol hexaphosphate rust-inhibiting release agent

Figure 2 Appearance of concrete after application of different demoulding agents to aluminium alloy formwork

4. Summary

- ZJTM embroidery-resistant release agent was prepared by using film-forming agent, compound emulsifier (HLB=9), corrosion inhibitor, co-emulsifier and stabilizer.
- The results show that the inositol hexaphosphate rust-inhibiting demoulding agent has short film-forming time and good stability, and forms a dense monomolecular protective film on the metal surface, thus slowing down the corrosion of the metal. After demoulding, the concrete has no harmful bubbles and no rust adherence.
- Inositol hexaphosphate rust-proof demoulding agent can prevent the reaction between aluminium sheet and concrete, and improve the ash and bubbles on the surface of concrete after using aluminium alloy formwork. Its application effect is better than other products in the market.

Reference

[1] Han Quanwei. Application of concrete demoulding agent and related engineering materials [M]. Beijing: China Building Materials Industry Press, 2008.

[2] Chen Xiaolu. Development of a water-borne demoulding agent for concrete [J]. New building materials, 2017, 7:24-26.

[3] Djelal, C.; Vanhove, Y.; Chambellan, D.; Brisset, P. Influenceof the Application method of release agents on thickness of mould oils[J], Materials and Structures, 2009,43(5): 687-698.

IOP Conf. Series: Materials Science and Engineering **587** (2019) 012004 doi:10.1088/1757-899X/587/1/012004

[4] Zhang Hongsheng.Application of non-toxic phytic acid in metal protection[J].Electroplating and finishing,,2000;22(1): 1-4.
[5] Zhu Lei; Yu Ping; Luo Yunbai Progress in Research and Application of Imidazoline Corrosion Inhibitor [J] Material Protection, 2003, 36 (12): 4-6.

Study on the application of embolization materials of collagen

Na Wang[1], Xue Jiang[2], Jiwei Wang[3] and Xiaomei Hao[4,*]

[1]Department of Anesthesiology, First Hospital of Jilin University, China
[2]Department of Anesthesiology, First Hospital of Jilin University, China
[3]The Third Operating Room, First Hospital of Jilin University, China
[4]Department of Archive Room, First Hospital of Jilin University, China

Corresponding author and e-mail: Xiaomei Hao, 2316246@qq.com

Abstract. The embolization mechanism, advantages and disadvantages as well as clinical application of collagen embolization materials were retrospectively analyzed, and the correlation between their application fields and the selection of embolization materials was discussed, so as to provide references for clinical treatment. Relevant literatures in Pubmed and Web of Science databases were searched, and the literature on clinical application and experimental research of embolic materials was systematically evaluated. Embolic materials should be biodegradable and biocompatible. Embolic materials for clinical use include absorbable gelatin sponge, polyvinyl alcohol particle embolization agent, micro-spring coil and liquid embolization materials. Gelatin embolization materials have the advantages of non-antigenicity, histocompatibility, convenient preparation, excellent compressibility and expansion in water. Endovascular embolization materials have been produced and applied in clinical practice, but new problems and new requirements have been rising. All kinds of embolic materials have advantages and disadvantages, and none of them can be suitable for all diseases. Therefore, it is very important to select suitable embolic materials for safe and effective treatment.

1. Introduction

In recent years, with the promotion of endovascular embolization technology, embolic materials play an important role in the application of endovascular embolization [1, 2]. However, all kinds of materials have their own advantages and disadvantages. With the development of biological materials, how to choose suitable embolic materials is the focus of the current research, which is very important for safe and effective treatment.

The properties and applications of various bioembolic materials will be reviewed. Endovascular embolism has the characteristics of small trauma, strong repeatability, accurate positioning, high efficacy, quick response, low complications, and the combined application of various technologies. Vascular repair through endovascular embolization is the best cure [3, 4]. Instruments used during embolization include catheters and guidewires. Catheters include angiographic catheter, microcatheter, such as microcatheter for intracranial artery, microcatheter for peripheral artery interventional therapy, etc. Guidewires used to guide the catheters insist in selective or super-selective intubation, which can effectively protect the vessel wall and catheter head from damage. In embolization therapy, guide wire is often used to guide the catheter to the target vessels for corresponding treatment. In order to achieve transcatheter injection and vascular embolization, appropriate micro catheter is needed to complete the embolization. In addition, the good characters of the material, killed operation of the doctor, video monitoring equipment are all factors affecting recanalization. However, embolism materials can be absorbed, but it also can cause serious damage to target if vascular embolization materials are not suitable. The choice of embolic materials is an important factor in embolization therapy [5].

2. Absorbable embolic material gelatin sponge

2.1. Advantages

The absorbable embolic material gelatin sponge is a biodegradable intermediate embolic material in vivo, which is not only a kind of collagen, but also a kind of macromolecular biological material [6]. Gelatin sponge is a spongy solid preparation made of gelatin liquid after foaming, curing, freeze-drying and sterilization. It has good histocompatibility, blood compatibility, biodegradability, weak antigenicity and biological safety requirements, making it an excellent embolization material [7]. It can be cut into small pieces according to the size of the target blood vessels, small strips or 1.0-2.0mm particles into 2-5 mL syringe, its characteristics are compressibility, reexpansibility, non-toxic no antigen, low price, simple preparation, easy to inject, reliable embolization, generally 1-3 weeks absorption. Collagen is the most abundant and widely distributed protein in animals. Collagen has unique triple helix structure, low antigenicity, non-toxicity, good biocompatibility and biodegradability, etc. [8-10]. Cellulose oxide ACTS like a gelatin sponge [11].

2.2. Disadvantages

At present, the collagen used in China mainly comes from the skin and Achilles tendon of pigs and cows. The preparation of gelatin sponge from the residual system of rana chensinensis can promote the comprehensive utilization of rana chensinensis and increase the added value of rana chensinensis [12]. Fish scales and skin about more than 40, ten thousand tons of waste of aquatic products, such as the waste of aquatic products are rich in collagen protein, its amino acid composition and terrestrial mammal collagen there was no significant difference, and it does not exist zoonotic infectious disease, development and utilization of great value, develop skin and other aquatic products waste using new approaches, not only can improve the added value of aquatic products processing, and can reduce the pollution of the environment, has the good economic and social benefits. Angiography was performed before the use of gelfoam to understand whether the blood supply artery and intubation route of the lesion have relevant collateral circulation [13].

3. How to select the suitable materials

Superselective intubation was performed according to the standard technique, and the position of catheter insertion should be as close as possible to the treatment site to prevent embolization of normal vessels. Particles of appropriate size are selected according to the condition of the lesion, and improper selection of particle size may lead to particles entering the blood supply artery of the normal tissue or entering the outflow vein of the lesion [14]. Appropriate amount of low concentration contrast agent (usually 30%-40%), ionic type such as dimeglutamine or non-ionic type such as iopromide, triiodotriamide hexanol benzene mixed with gelatin sponge particles make a uniform suspension with appropriate concentration, there shall be no foam. The suspension is inhaled into the syringe with appropriate concentration and fine particle suspension. If the suspension is too thick, the catheter can be blocked and it should be avoided.

3.1. Possible interactions with other materials or drugs

Under angiographic fluoroscopy, gelatin sponge particle suspension was injected through the catheter at an appropriate rate without reflux, otherwise it might enter the non-embolic artery and cause ectopic embolism, thus leading to tissue and organ necrosis [15]. The gelatin sponge particles were continuously injected until the results were satisfactory. Gelatin sponge is one of the most widely used embolization materials at present, which is suitable for embolization treatment of various hemorrhagic tumors and certain vascular malformations due to its safety, non-toxicity, easy sampling and convenient use [16]. However, as biomedical materials, aquaculture collagen and terrestrial mammal collagen have common disadvantages, such as fast degradation in vivo and poor flexibility, which limit its application.

3.2. Programmed preparation

Glutaraldehyde as a crosslinking agent can also increase the mechanical strength of collagen medical materials and slow down its degradation rate in vivo. Chitosan was added to gelatin to make a sponge preparation that can stop bleeding, promote the growth of granulation tissue and accelerate wound healing [17]. However, it has its own disadvantages.

It is non-permanent occlusion, and usually begins to be absorbed within 14-19 days after vascular embolism. It may be completely absorbed after 3 months, and it is easy to form recanalization. In addition, gelfoam should not be used in patients with difficulty in super-selective intubation, small arterial particles, larger arteriovenous fistula than particles, and patients requiring permanent embolization, etc. [18].

Stent implantation is a common way to treat patients with cardiovascular diseases, which are serious threats to human health. Conventional metal stents are mostly used in the past, and there are some hidden dangers for the metal stents that remain in the blood vessels when they are effectively treated [19]. The blood vessels will contract and relax regularly. If the blood vessels are always supported by the scaffold, the blood vessels will become stiff and lose elasticity, leading to various adverse reactions. For example, the vessel at the stent site lost its elasticity, vascular restenosis, thrombosis, etc., and affected the patient's subsequent surgical bypass [20]. In recent years, new stent has been developed. The advantage of the new scaffolds is that they are fully biodegradable and are mainly made of naturally biodegradable materials. After the implantation of blood vessels into the human body, after the treated blood vessels have healed and no longer need the support of the scaffold, the scaffold can be continuously degraded over time and eventually disappear completely. Magnesium and its alloys have excellent mechanical properties and biocompatibility [21]. The scaffold has the unique advantages of strength of metal scaffold and bioabsorbability of degradable polymer scaffold. Due to the special microenvironment inside the blood vessels, the stent should have good corrosion resistance.

4. Conclusions

The first generation of blood compatibility materials has experienced a history of more than half a century, and so far it has not been able to truly meet the clinical requirements. A variety of factors affect the blood compatibility, because of the complex changes of coagulation mechanism, material surface characteristics. How to modify the material become an important way to improve the blood compatibility. Currently, there are still some problems in the existing evaluation system. How to select effective materials and how to quantify the evaluation results have become important directions for improving blood compatibility.

Acknowledgement

This research was financially supported by the First Hospital of Jilin University.

References

[1] Nguyen T, Kalish J, Woodson J. Management of civilian and military vascular trauma: lessons learned. Semin Vasc Surg. 2010;23(4):235-242.

[2] Tsoumakidou G, Theocharis S, Ptohis N, et al. Liver hypertrophy after percutaneous portal vein embolization: comparison of N-butyl-2-cyanocrylate versus sodium acrylate-vinyl alcohol copolymer particles in a swine model. Cardiovasc Intervent Radiol. 2011;34(5):1042-1049.

[3] Siddique MK, Majeed S, Irfan M, et al. Missed vascular injuries: presentation and outcome. J Coll Physicians Surg Pak. 2014;24(6):428-431.

[4] Goksu E, Yuruktumen A, Kaya H. Traumatic pseudoaneurysm and arteriovenous fistula detected by bedside ultrasound. J Emerg Med. 2014;46(5):667-669.

[5] Vanhegan IS, Shehzad KN, Bhatti TS, et al. Acute popliteal pseudoaneurysm rupture secondary to distal femoral osteochondroma in a patient with hereditary multiple exostoses. Ann R Coll Surg Engl. 2012;94(3):e134-136.

[6] Yuki I, Uchiyama N, Murayama Y, et al. Intravascular tissue reactions induced by various types of bioabsorbable polymeric materials: correlation between the degradation profiles and corresponding tissue reactions. Neuroradiology. 2010;52(11):1017-1024.

[7] Calton WC Jr, Franklin DP, Elmore JR, et al. Ultrasound-guided thrombin injection is a safe and durable treatment for femoral pseudoaneurysms. Vasc Surg. 2001; 35(5):379-383.

[8] Kubo M, Kuwayama N, Hirashima Y, et al. Hydroxyapatite ceramics as a particulate embolic material: report of the physical properties of the hydroxyapatite particles and the animal study. AJNR Am J Neuroradiol. 2003;24(8): 1540-1544.

[9] Spigos DG, Jonasson O, Mozes M, et al. Partial splenic embolization in the treament of hypersplenism. AJR Am J Roentgenol. 1979;132(5):777-782.

[10] Shi GG, Yao SG, Wang ZD, et al. Clinical research on delayed traumatic epistaxis and pseudoaneurysm. Zhonghua Er Bi Yan Hou Tou Jing Wai Ke Za Zhi. 2008;43(6):414-418.

[11] Oh JS, Lee HG, Chun HJ, et al. Evaluation of arterial impairment after experimental gelatin sponge embolization in a rabbit renal model. Korean J Radiol. 2015;16(1):133-138.

[12] Aksoy M, Taviloglu K, Yanar H, et al. Percutaneous transcatheter embolization in arterial injuries of the lower limbs. Acta Radiol. 2005;46(5):471-475.

[13] Dabus G, Pizzolato R, Lin E, et al. Endovascular treatment for traumatic scalp arteriovenous fistulas: results with Onyx embolization. J Neurointerv Surg. 2014;6(5):405-408.

[14] Kawarada O, Yokoi Y, Izuta M. Distal embolization during intracranial carotid artery stenting. Catheter Cardiovasc Interv. 2008;71(6):740-743.

[15] Kawarada O, Yokoi Y, Takemoto K. The characteristics of dissemination of embolic materials during renal artery stenting. Catheter Cardiovasc Interv. 2007;70(6):784-788.

[16] Matsumoto H, Terada T, Tsuura M, et al. Basic fibroblast growth factor released from a platinum coil with a polyvinyl alcohol core enhances cellular proliferation and vascular wall thickness: an in vitro and in vivo study. Neurosurgery. 2003;53(2):402-407.

[17] Swanson KE, Bartholomew JR, Paulson R.Hypothenar hammer syndrome: a case and brief review. Vasc Med. 2012 Apr;17(2):108-15.

[18] Bravman JT, Ipaktchi K, Biffl WL, Stahel PF.Vascular injuries after minor blunt upper extremity trauma: pitfalls in the recognition and diagnosis of potential "near miss" injuries. Scand J Trauma Resusc Emerg Med. 2008 Nov 25;16:16. doi: 10.1186/1757-7241-16-16.

[19] Wu J, Perron AD, Miller MD, Powell SM, Brady WJ. Orthopedic pitfalls in the ED: pediatric supracondylar humerus fractures. Am J Emerg Med. 2002 Oct;20(6):544-50.

[20] Bibbo C, Lin SS, Cunningham FJ. Acute traumatic compartment syndrome of the foot in children. Pediatr Emerg Care. 2000 Aug;16(4):244-8.

[21] Martella AT, Coomaraswamy M.Delayed presentation of a penetrating suprarenal aortic injury: case report. J Trauma. 1993 Jan;34(1):148-50.

IOP Conf. Series: Materials Science and Engineering **587** (2019) 012006 doi:10.1088/1757-899X/587/1/012006

Study on the application of embolization materials of polyvinyl alcohol particles

Na Wang[1], Jialin Gao[2], Yu Wu[3] and Wei Han[4,*]

[1]Department of Anesthesiology, First Hospital of Jilin University, China
[2]Department of Urology, First Hospital of Jilin University, China
[3]Department of Otorhinolaryngology, First Hospital of Jilin University, China
[4]Department of Anesthesiology, First Hospital of Jilin University, China

Corresponding author and e-mail: Wei Han, 2212592021@qq.com

Abstract. The application, advantages and disadvantages of PVA particle embolization materials were analyzed retrospectively, and the correlation between the application field and the choice of embolization materials was discussed, which provided a reference for clinical treatment. Relevant literatures in PubMed and Web of Science databases were searched, and the literature on clinical application and experimental research of embolic materials was systematically evaluated. The embolization mechanism, advantages and disadvantages, and clinical application of PVA particle embolization materials were discussed. Polyvinyl alcohol particles are such kind of polymer material that insoluble in water, high expansion coefficient, mechanical embolization of blood vessels, no sensitization reaction, no systemic acute toxicity, no intradermal stimulation reaction, safe and effective use. The adhesive liquid embolism material has no vascular toxicity and has been widely used. The application and upgrading of thrombolytic surface materials can significantly improve the blood compatibility of the materials. All kinds of embolic materials have their own advantages and disadvantages, and none of them can be suitable for all diseases. Therefore, it is very important to choose suitable embolic materials for safe and effective treatment.

1. Introduction

Excellent embolic materials should be non-toxic, non-antigenicity, biocompatible, easy to obtain and disinfect, not through X-ray and easy to be injected through catheter. In the process of targeted interventional therapy, the target blood vessels can be completely blocked, the abnormal blood flow channels in the body can be stopped quickly, and the tissue blood flow can be redistributed to ensure the efficacy [1]. It can control the time of occlusion of blood vessels. Once required, percutaneous recovery or recanalization of blood vessels are higher requirements for bioembolic materials [2]. Embolic material of a lot of kinds, according to the nature can be divided into solid and liquid, according to the body can be absorbed in the body can be divided into two types of absorption and non-absorption. At present, gelfoam, polyvinyl alcohol, anhydrous ethanol, isobutyl-2-cyanoacrylate, n-butyl-2-cyanoacrylate are widely used. Gelfoam can be disinfected without antigenicity and can be made into granules or strips of different sizes according to the needs. It is convenient to prepare and can occlude blood vessels from several weeks to several months. Polyvinyl alcohol is a synthetic material with the characteristics of long occlusion time, non-absorption, hygroscopic expansion and so on.

Isobutyl -2- cyanacrylate is a liquid tissue adhesive, in the case of ionic substances, such as blood and ion type contrast agent soon after polymerization curing, can long-term occlusion of blood vessels, often used in arteriovenous malformations, esophageal varices bleeding. The polymerization time was delayed by adding appropriate amount of iodized oil and iodized benzyl ester, and made it opaque to X-ray. The similar preparation is n-butyl-2-cyanoacrylate, which has a long polymerization time and is beneficial to technical operation. Recently, it has gradually replaced isobutyl-2-cyanoacrylate for clinical use [3]. Autologous blood clot is the only short-term embolism at present. The embolization time is 24-48 h. The advantages of autologous blood clot are convenient sampling and no antigenicity. Spring coil for mechanical emboli can be used for large and small arteries, permanent occlusion of blood vessels, no active role in the body [4].

2. Advantages and disadvantages of polyvinyl alcohol particles

2.1. Advantages

Polyvinyl alcohol particles are permanent embolizers, a kind of polymer material, white or yellowish-light and soft porous sponge particles and porous irregular particles. The raw material is polyvinyl alcohol, which must be mixed with the contrast agent before development [5]. When used, polyvinyl alcohol needs to be cut into pieces to soak with contrast agent, which is suitable for embolization of large and medium-sized blood vessels. Powder is used for embolization of peripheral blood vessels, non-toxic, insoluble in water, with good histopcompatibility and high expansion coefficient. Fibrous tissue can grow into it quickly and mechanically block the blood vessels after being injected into the blood vessel system. Polyvinyl alcohol particles foam embolization particles of different specifications were designed for embolization of vascular-rich tumors and intravascular embolization of arteriovenous malformations according to particle diameter and size [6].

2.2. Disadvantages

Polyvinyl alcohol particles were used for embolization of tumor arteries or malformed vessels. Polyvinyl alcohol particles granules were injected into human body to mechanically embolize vessels at the lesion site. Thrombus materials and mechanism were formed in the gap between polyvinyl alcohol particles granules to permanently embolize vessels. Polyvinyl alcohol particles injected into the human body are treated by mechanical embolization of blood vessels, which has no pharmacological effect on its own. Insoluble in water and ethanol, soluble in formic acid, with good biological safety, mechanical embolization of blood vessels, no sensitization, no systemic acute toxicity, no intradermal stimulation, safe and effective use [7].

Angiography was performed before use to understand the blood supply artery and intubation pathway of the lesion and the related collateral circulation [8]. Superselective intubation was performed according to the standard technique, and the position of catheter insertion should be as close as possible to the treatment site to prevent embolization of normal vessels. Polyvinyl alcohol particles of appropriate size were selected according to the condition of the lesion, and improper selection of particle size may lead to particles entering the blood supply vein of normal tissue or entering the outflow vein of the lesion [9-12].

3. Compared with other materials

3.1. Spring coils

The spring coil which has good compliance and accurate embolization positioning can fully fill the gap beyond the reach of the balloon and smaller fistula, and can complete the embolization of larger diameter through a thinner catheter [13]. At the same time, the spring coil has a relatively high safety in the operation process, does not cause damage to the vascular intima, and has good controllability, randomness and permanent embolization. It can be shaped according to the size of aneurysm and morphology, and different types of spring coil can be selected according to the size of the fistula [14]. During embolization, the coil was placed in the aneurysm to promote intra aneurysm thrombosis. Even if the coil was improperly placed, it was recovered and replaced, which could be observed by X-ray plain film long-term follow-up. In addition, the application of spring coil embolization of fistula can achieve dense embolization and reduce the recurrence rate [15]. Its disadvantages lie in its incomplete embolization, which is a proximal embolization material, easy to establish collateral circulation, high friction in microcatheter, poor biocompatibility, displacement of spring coil and embolization of normal blood vessels, so it is not an ideal embolization material [16].

3.2. Surface materials

The main anticoagulant heparin is usually fixed on the surface of the polymer, which can significantly improve the blood compatibility of the material [17]. Heparin is a natural anticoagulant with strong anticoagulant effect [18]. The strong anionic properties of heparin are combined with polymer materials with cationic surfaces to form heparinized hydrophilic polymer materials. If the material continues to release heparin in contact with blood, local thrombosis can be prevented [19]. The fixation of heparin on the material surface includes physical adsorption and chemical binding, and in the chemical binding, there are ionic bond binding and covalent bond binding. Heparin fixed by chemical bonds is more stable than that by physical adsorption, while that by covalent bonds is much more stable than that by ionic bonds in chemical fixation. On the other hand, the biological activity of heparin immobilized on the surface through covalent binding is low, so there is a problem that the size of improved anticoagulant is not very prominent [20].

In order to ensure the treatment effect and improve the safety of treatment, the scaffold materials used in clinical should have good biocompatibility. Specifically, it mainly involves blood compatibility and cell compatibility [21]. In the application process of various biological materials, good anticoagulant energy should be satisfied, so as to help reduce the tendency of thrombosis and avoid the occurrence of various risk events in clinical treatment. Platelet adhesion assay is the most direct and commonly used method to evaluate and judge the anticoagulant properties of blood contact materials [22]. By observing the adhesion of platelets on the surface of the material, we can intuitively understand the blood compatibility of the material. In the experiment of platelet adhesion, a small number of platelets could be observed on the surface of the material without aggregation by electron microscopy scanning and fluorescence staining [23].

4. Conclusions

At present, improving the blood compatibility of biomaterials is a crucial step in the research of biomaterials, and the surface modification of biomaterials is the key of this step. However, when grafting materials on the surface, the requirements of blood compatibility of grafted groups or substances should be considered, and the strength of the bond between grafted groups and matrix materials should also be considered. Usually, the ideal modification effect can be achieved by coupling the grafted materials with appropriate intermediate media. There are many methods for the preparation of blood compatible biomaterials. The optimization and modification of material surface interface characteristics have become an important way to improve its blood compatibility, but there are still some problems in the existing evaluation system, such as the selection of sensitive indicators and effective positive control materials and the quantification of evaluation results.

Acknowledgement

This research was financially supported by the First Hospital of Jilin University.

References

[1] Falagas ME, Karavasiou AI, Bliziotis IA. A bibliometric analysis of global trends of research productivity in tropical medicine. Acta Trop. 2006; 99(2-3): 155-159.

[2] Kumari L, Li W, Huang JY, et al. Solvothermal Synthesis, Structure and Optical Property of Nanosized CoSb(3) kutterudite. Nanoscale Res Lett. 2010; 5(10): 1698-1705.
[3] Ye JY, Huang ZY, Min C, et al. Structure and haemocompatibility of ZnO films deposited by radio frequency sputtering. Biomed Mater. 2009; 4(5): 055004.
[4] Rodrigues SN, Goncalves IC, Martins MC. Fibrinogen adsorption， platelet adhesion and activation on mixed hydroxyl methylterminated self-assembled monolayers. Biomaterials. 2006; 27(31): 5357-5367.
[5] Bondar B, Fuchs S, Motta A, et al. Functionality of endothelialcells on silk fibroin nets: comparative study of micro- andnanometric fibre size. Biomaterials. 2008; 29(5): 561-572.
[6] de Mel A, Jell G, Stevens MM, et al. Biofunctionalization of biomaterials for accelerated in situ endothelialization: a review. Biomacromolecules. 2008; 9(11): 2969-2979.
[7] Rodney AW. Vascular prothesis present status and furture. Development[z].1989; 7(6): 443-449.
[8] van Det RJ, Vriens BHR, van der Palen J, et al. Dacron or PTFE for femoro-popliteal above-knee bypass grafting: short-and long-term results of a multicentre randomized trial. Eur J Vasc Endovasc Surg. 2009; 37(4): 457-463.
[9] van Det RJ, Vriens BHR, van der Palen J, et al. Dacron or PTFE for femoro-popliteal above-knee bypass grafting: short-and long-term results of a multicentre randomized trial. Eur J Vasc Endovasc Surg.2009;37(4):457-463.
[10] Soffer L, Wang X, Zhang X, et al. Silk-based electrospun tubularscaffolds for tissue-engineered vascular grafts. J Biomater SciPolymEd. 2008; 19(5): 653-664.
[11] Shao Z, Vollrath F. Surprising strength of silkworm silk. Nature. 2002; 418(6899): 741.
[12] Jin HJ, Kaplan DL. Mechanism of silk processing in insects and spiders. Nature. 2003; 424(6952): 1057-1061.
[13] Sperling C, Fischer M, Maitz MF, et al. Blood coagulation on biomaterials requires the combination of distinct activation processes. Biomaterials. 2009; 30(27): 4447-4456.
[14] Valluzzi R, Gido SP, Zhang W, et al. Trigonal crystal structure of Bombyx mori silk incorporating a threefold helical chain conformation found at the air-water interface. Macromolecules. 1996; 29(27): 8606-8614.
[15] Valluzzi R, He SJ, Gido SP, et al. Bombyx mori silk fibroin liquid crystallinity and crystallization at aqueous fibroin–organic solvent interfaces.Int J Biol Macromol. 1999; 24(2-3): 227-236.
[16] Majumdar P, Webster DC. Preparation of siloxane-urethane coatings having spontaneously formed stable biphasic microtopograpicalsurfaces. Macromolecules. 2005; 38(14): 5857-5867.
[17] Antonsen L, Maeng M, Thayssen P. Intimal hyperplasia and vascular remodeling after everolimus-eluting and sirolimus-eluting stent implantation in diabetic patients: the randomized Diabetes and Drug-Eluting Stent (DiabeDES) IV Intravascular Ultrasound trial. Catheter Cardiovasc Interv. 2014: 83(6): 864-872.
[18] Phang ZH, Miskon MFB, Ibrahim SB. Blunt trauma to the antecubital fossa causing brachial artery injury and minor fractures around the elbow joint, an easily missed diagnosis with potential devastating consequences: a case report. J Med Case Rep. 2018 Jul 18;12(1):211. doi: 10.1186/s13256-018-1751-7.
[19] Siddique MK, Majeed S, Irfan M, Ahmad N. Missed vascular injuries: presentation and outcome. J Coll Physicians Surg Pak. 2014 Jun;24(6):428-31. doi: 06.2014/JCPSP.428431.
[20] Siddique MK, Bhatti AM. A two-year experience of treating vascular trauma in the extremities in a military hospital. J Pak Med Assoc. 2013 Mar;63(3):327-30.
[21] Pandya NK, Edmonds EW, Roocroft JH, Mubarak SJ.Tibial tubercle fractures: complications, classification, and the need for intra-articular assessment. J Pediatr Orthop. 2012 Dec;32(8):749-59. doi: 10.1097/BPO.0b013e318271bb05.
[22] Bartkowiak-Jowsa M，B dzi6ski R，Koztowska Anna. Mechanical，rheological，fatigue，

IOP Conf. Series: Materials Science and Engineering **587** (2019) 012006 doi:10.1088/1757-899X/587/1/012006

and degradation behavior of PLLA, PGLA and PDGI_A as materials for vascular implants. Meccanica. 2013; 48(3): 721-731.

[23] Hlawaty H, Jacob MP, Louedec L. Leukotriene receptor antagonism and the prevention of extracellular matrix degradation during atherosclerosis and in-stent stenosis. Arterioscler Thromb Vasc Bi01. 2009; 29(4): 518-524.

IOP Conf. Series: Materials Science and Engineering **587** (2019) 012007 doi:10.1088/1757-899X/587/1/012007

Investigation on the water resistance of the fly-ash modified magnesium phosphate cement

Sihui Sun[1], Runqing Liu[1], Xingke Zhao[1*], Yao Zhang[2], Yuanquan Yang[3]
[1]Shenyang Ligong University, Nanping RD 6, New Hunnan District, Shenyang, China, 110159
[2]China Railway Construction Bridge Engineering Bureau Group CO.,LTD, Dalian, China, 116021
[3]Faculty of Infrastructure Engineering Dalian University of Technology, Dalian, China,116024

Corresponding author and e-mail: Zhao Xingke, 157371978@qq.com

Abstract: In order to improve the water resistance of magnesium phosphate cement (MPC), the flexural strength, compressive strength, strength retention rate and mass retention rate under different immersion conditions, were tested by XRD and SEM methods by adding 30%, 35%, 40%, 45% and 50% fly ash. The mechanism of fly ash improving the water resistance of magnesium phosphate cement was analyzed. The results show that when fly ash content is 35%, the mechanical properties of magnesium phosphate cement (MPC) reach the optimal, and the strength retention rate and mass retention rate are improved. Fly ash modified by magnesium phosphate cement produces no new hydration products. After it is soaked in water, a large number of flake like k-struvite are produced on the surface of MgO particles. Few cracks forms on the surface, which reduces the porosity and enhances the compactness, thus improving the water resistance of MPC cement.

1. Introduction

Magnesium phosphate cement has been discovered and used as early as 1939, which is made of MgO, phosphate, retarder, admixture and so on. Magnesium phosphate cement is different from ordinary Portland cement, which is mainly composed of three kinds of materials. Its preparation method is simple, preparation process is low energy consumption as well as is low water demand. Magnesium phosphate cement has the advantages of rapid setting and hardening, early strength. It can well meet the requirements of urgent repair and construction in wartime. Therefore, magnesium phosphate cement has a broad application prospect in military engineering. With the development of research on magnesium phosphate cement, the excellent performance has been recognized by the academic community. In recent years, there has been further development in the fields of biomedicine, oil well solidification, waste solidification, radioactive wastes solidification, 3D printing technology, etc. However, magnesium phosphate cement's strength obviously decreases in the humid environment or in contact with water, which will inevitably have an impact on its durability. Therefore, the study on the water resistance of modified magnesium phosphate cement has become one of the hot topics in this field.

Fly ash, as an admixture, has been widely used in building materials. Adding fly ash to magnesium phosphate cement can not only change the color of magnesium phosphate cement but also make it similar to the color of common cement pavement. It can also improve the performance of magnesium phosphate cement paste and mortar. Liu Kai believes that the appropriate addition of fly ash can improve the strength of magnesium phosphate cement. Wang Hongtao believes that fine fly ash, on the one hand, can extend the setting time of magnesium phosphate cement and improve its fluidity. Under the condition of maintaining the fluidity, it can reduce the water-solid ratio and improve the strength. Zhang Siyu studied the influence of fly ash on the expansion performance of magnesium phosphate cement and found that the expansion rate was the highest when the fly ash content was 30%. Ding Zhu believes that when fly ash content is 40%, the strength of magnesium phosphate cement reached the maximum. The results show that fly ash has an active effect, micro-aggregate effect, morphological effect and adsorption effect in magnesium phosphate cement. This paper mainly studies the influence of different fly ash content on the water resistance of magnesium phosphate cement, through the test and analysis of macro strength and micro performance, and the mechanism of improving the water resistance of magnesium phosphate cement by fly ash is further discussed.

2. Experimental details and methodology

2.1. Materials

Dead burned magnesia was prepared by magnesite (mainly including $MgCO_3$) burnt at about 1500℃ for 4 hours, then crushed and ground into particles with different sizes. The chemical composition the MgO was shown in Table 1.

Table 1. Chemical composition of the magnesia/%

oxide	MgO	SiO_2	CaO	Al_2O_3	Fe_2O_3
content /%	92	0.7	1.36	0.3	0.3

KH_2PO_4, abbreviated as P, is produced by Tianjin Binhai Fuli chemical plant with a content of 98.56%. Chemical composition of the $KH_2PO_4$is shown in Table 2.

Table 2. Chemical composition of potassium dihydrogen phosphate /%

composition	KH_2PO_4	water	Water insoluble substance	chloride	Fe	As	Heavy metal	KCl	Ph
content	98.56	2.12	0.04	0.1	0.001	0.001	0.001	34	4.52

In this work, borax was used as retarder. Its chemical formula is $Na_2B_4O_7$•10H2O, chemically pure, abbreviated as B. The content is no less than 99.4%. The chemical composition of borax is shown in Table 3.

Table 3. Chemical composition of borax /%

composition	$Na_2B_4O_7\cdot 10H_2O$	carbonate	Water insoluble substance	sulfate	Fe
content	95	0.2	0.002	0.2	0.005

Grade II fly ash (F) is chosen , which belongs to the low calcium ash, its specific surface area is about 280 m^2 / kg, the chemical composition of fly ash analysis as shown in Table 4.

Table 4. Chemical composition of fly ash /%

composition	SiO_2	Al_2O_3	Fe_2O_3	CaO	MgO	TiO_2	SO_3	Loss
content	39.5	26.8	15.5	4.1	1.5	2.8	0.6	8.4

The sand used in the experiment is ISO standard sand.

The test water is ordinary tap water.

2.2. Sample preparation

The optimal mixture ratio was selected through the preliminary foundation test. The magnesium to phosphorus by molar ratio is 4:1. Borax was 16% of the magnesia. The fly ash is 30%, 35%, 40%, 45%, and 50% replacement of the sum mass of magnesia and KH_2PO_4, respectively. Raw materials were weighted and poured into the stirring container for 2min at low speed. After that water was added and the mixture was stirred quickly at high speed for 1min. The stirred magnesium potassium phosphate cement slurry was rapidly poured into the mold with size of 40mm×40mm×160mm sample. The sample was vibrated for 1min30s to make the mixture compacted. The sample was removed from the mold within 30min~1h after the sample hardened.

2.3. Experimental methods

2.3.1. Water resistance

In this paper, the strength retention rate is used to evaluate the water resistance of materials. According to the strength loss, the strength retention rate K value (accurate to 0.01) was proposed as the index evaluation of the water resistance of the MPC specimen. The strength retention K value is defined as follows:

$$K_t = \frac{f}{F} \times 100\% \quad (1)$$

Kt -- strength retention rate of MPC specimen for t days;

F -- unconfined compressive strength, MPa, of hydrotropic specimens;

F -- unconfined compressive strength of natural curing of MPC specimens, MPa.

When the value of strength retention rate K is small, it means that the strength loss of magnesium phosphate cement after immersion in water is small, it means that the water resistance of magnesium phosphate cement is well at this time, and the water resistance is poor if the value of K is high. Therefore, the strength retention rate can directly characterize the water resistance of magnesium phosphate cement in the same system.

Softening coefficient, as the traditional evaluation index of resistance to water, was not used, due to that in the softening coefficient test (according to the standard JGJ51-2002), compressive strength of the specimens need to be dried to constant weight under 105~110°C in advance, and the main hydration products of magnesium phosphate cement $MgKPO_4 \cdot 6H_2O$(MKP) can however decomposed in the temperature of 60~70°C. Thus softening coefficient was not used here as an evaluation index for water resistance of magnesium phosphate cement.

2.3.2. SEM &XRD Text:

Ultima IV X-ray diffractometer manufactured by Japan Rigaku company was used for phase identification. The microstructure of S-3400N microcomputer electron scanning microscope manufactured by German company HITACHI was used to perform microstructure test.

3. Results and discussion

The flexural strength and compressive strength development of magnesium potassium phosphate cement samples with different content of fly ash, which were immersed in static water (changing water every 8 hours), were tested. The strength retention rate and mass retention rate (see Table 6 for specific values) were calculated to characterize the water resistance of the magnesium potassium phosphate cement. The mechanism of water resistance of magnesium phosphate cement modified by fly ash is discussed through phase identification and morphology analysis, which provides theoretical basis and scientific guidance for the engineering practice of long-term contact between magnesium phosphate cement and water.

Table 5. Flexural and compressive strength of magnesium phosphate cement with fly ash

fly ash	Flexural strength /MPa					The compressive strength /MPa				
	30%	35%	40%	45%	50%	30%	35%	40%	45%	50%
8d	8.4	8.6	7.8	7.4	7.3	32.7	33.5	29.3	24.7	23.3
7d+1d*	8.2	8.4	7.7	7.3	7.1	31.3	32.3	28.7	23.0	22.8
7d+1d**	8.1	8.3	7.6	7.1	6.9	30.7	31.8	26.5	21.6	20.6
14d	9.1	9.2	9.0	8.9	8.8	36.1	37.1	36.5	33.3	32.8
7d+7d*	8.7	8.9	8.5	8.3	8.1	33.3	34.8	32.3	29.1	28.3
7d+7d**	8.5	8.8	8.4	8.2	7.9	32.2	33.9	31.8	27.8	27.1

Note: * means no water immersion, ** means water immersion every 8 hours, and no * means curing in air

Table 6. Strength and quality retention of magnesium phosphate cement with fly ash

fly ash	Strength retention /%					Quality retention /%				
	30%	35%	40%	45%	50%	30%	35%	40%	45%	50%
7d+1d*	95.7	96.4	97.9	93.1	89.1	—	—	—	—	—
7d+1d**	98.3	94.9	90.4	87.4	88.4	—	—	—	—	—
7d+7d*	92.2	93.9	88.4	87.3	86.3	99.6	99.7	99.7	99.3	99.6
7d+7d**	88.6	91.5	87.1	83.4	82.6	99.4	99.4	99.3	98.6	99.1

Note: * means no water immersion, and ** means water immersion every 8 hours

3.1. Influence of fly ash content on mechanical properties of magnesium phosphate cement

3.1.1. Influence of fly ash content on flexural strength of magnesium phosphate cement

Magnesium phosphate cement with different fly ash content is cured in the air for 8d. After curing in the air for 7d, then it was soaked in static water for 1d*. The specimens cured in the air for 7 days, soaked in water (every 8 hours) for 1d**, is shown in Table 1. The specimens cured in the air for 14 days and 7 days, then soaking in static water for 7d* and in air for 7 days, then soaked in water for 7** days, is shown in Figure 2.

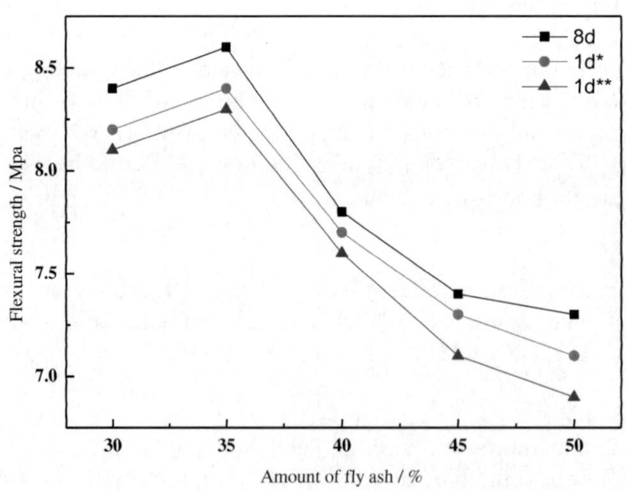

Figure 1. Effect of fly ash content on flexural strength of magnesium phosphate cement in different soaking conditions(soaking for one day)

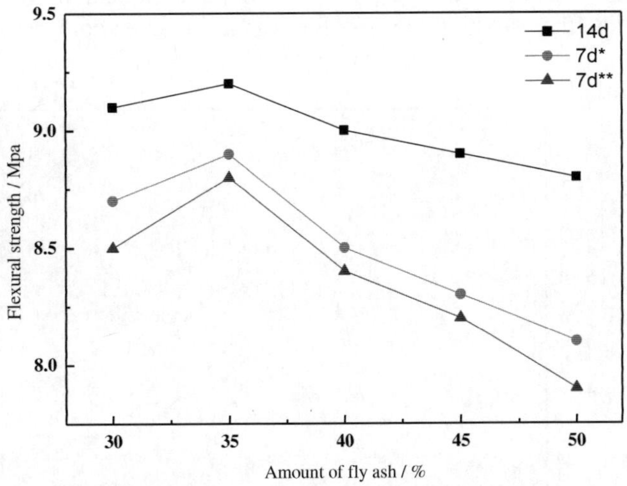

Figure 2. Effect of fly ash content on flexural strength of magnesium phosphate cement in different soaking conditions(soaking for seven days)

Fig.1 presents the flexural strength of fly ash containing magnesium phosphate cement soaked in water for 1d* and 1d**. The flexural strength of the MPC cured in the air in the three curves in the Figure is better than that soaked in water for 1d*on the whole, while the flexural strength of the MPC soaked in static water is better than that soaked in water every 8 hours. This indicates that the flexural strength of MPC adulterated with fly ash will shrink in water, and the flexural strength of MPC will shrink more in the dynamic water environment. The flexural strength of MPC increased first and then decreased with the increase of fly ash content. The group with good flexural strength shown in the Figure is magnesium phosphate cement with a fly ash content of 35%. The yield of this group has the best flexural strength whether maintained in the air for 8d (8.6mpa) or in static water for 1d*(8.4mpa) and changed water for 1d**(8.3mpa). The flexural strength of MPC with fly ash content of 50% is the lowest. As shown in Figure 2, the flexural strength of magnesium phosphate cement with different fly ash contents soaked for 7d* and 7d**, and the flexural strength trends of MPC curing in the air is the same as that in Figure 1. The flexural strength of the MPC cured in the air is higher than that of the MPC soaked for 7*days, while the flexural strength of the MPC soaked in static water is better than that of the MPC soaked in water changed every 8 hours. Moreover, the flexural strength after soaking in water for 7d decreased more than that after curing in air. The flexural strength of MPC increases first and then decreases with the increase of fly ash content. When the peak value of 8d curing in the air is 35%, it is 14d(9.2MPa), 7d*(8.9MPa) and 7d**(8.8MPa), respectively. The variation trend of the flexural strength of each fly ash content at the soaking 7d* and 7d** is similar to that in Figure 1. As shown in Figure 1, the MPC with a fly ash content of 50% has the lowest flexural strength.

When the fly ash content is 35%, the flexural strength of soaked MPC is higher than that of other content, indicating that when the fly ash content is 35%, the micro-aggregate effect of fly ash in magnesium phosphate cement reaches the maximum. The fly ash particles are evenly distributed in the cement slurry, reducing the porosity and increasing the compactness of the cement slurry. Magnesium phosphate cement slurry is relatively viscous. Adding fly ash can enhance its fluidity and promote the expulsion of bubbles in the slurry to increase the compactness. However, if the fly ash content is too large, a large amount of unreacted fly ash will lead to a decrease in strength.

3.1.2. Influence of fly ash content on compressive strength of magnesium phosphate cement

As shown in Figure 3, magnesium phosphate cement with different content of fly ash is cured in air for 8d, after curing with in air for 7d, soak in static water for 1d* and after curing in air for 7d, change

IOP Conf. Series: Materials Science and Engineering **587** (2019) 012007 doi:10.1088/1757-899X/587/1/012007

water (every 8 hours) to soak for 1d** the change in compressive strength. As shown in Figure 4, it is maintained in the air for 14 days, curing in the air for 7d and then soaking in static water for 7d* and curing in the air for 7d and then soaking in water for 7d** changes in compressive strength.

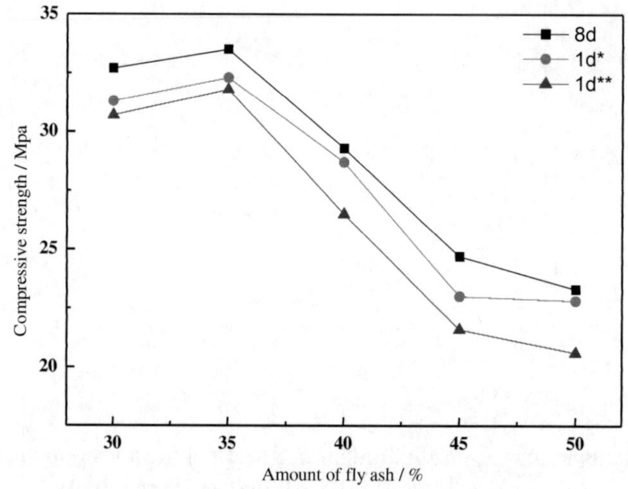

Figure 3. Effect of fly ash content on compressive strength of magnesium phosphate cement in different soaking conditions(soaking for one day)

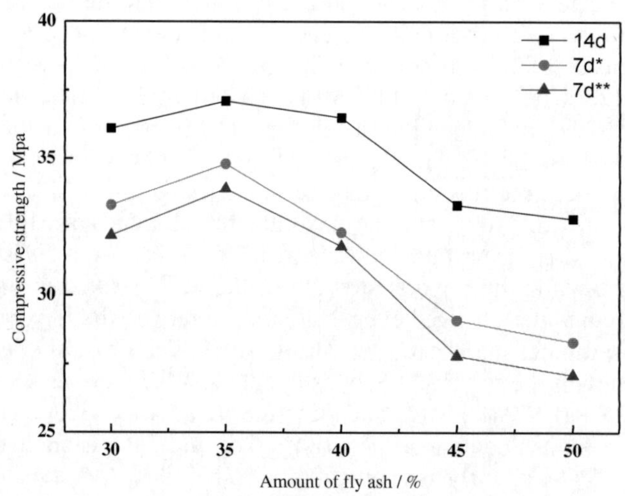

Figure 4. Effect of fly ash content on compressive strength of magnesium phosphate cement in different soaking conditions(soaking for seven days)

As shown in Figure 3, the compressive strength of magnesium phosphate cement soaked in water for 1d* and 1d** with different fly ash content. The general trend of the flexural strength development is quite similar to that of flexural strength in Figure 1 and Figure 2. The compressive strength of MPC cured in the air is generally better than that soaked in unchanged water for 1d*, while the compressive strength of MPC soaked in static water is better than that of MPC changed in water every 8 hours. This indicates that the compressive strength of MPC adulterated with fly ash will shrink in water, and the compressive strength of MPC will shrink more in a dynamic water environment. The compressive strength of MPC increased first and then decreased with the change of fly ash content. When the fly ash content is 35%, the compressive strength of this group is relatively good, whether it is maintained in the air for 8d(33.5MPa) or in static water for 1d*(32.3MPa) and changed water for 1d**(31.8MPa). The compressive strength of MPC with 50% fly ash content is the lowest. As shown in Figure 4, the compressive strength of different fly ash contents on magnesium phosphate cement soaked in water for 7d* and 7d**. The trends of the three compressive strength curves are the same as that of the curve shown in Figure 3. The compressive strength of MPC cured in the air is generally better than that soaked in water for 7 days, while the compressive strength of MPC soaked in static water is better than that soaked in water every 8 hours. Moreover, the compressive strength after soaking in water for 7d decreased significantly compared with that after curing in air. The compressive strength with 35% of fly ash at 14d, 7d* and 7d* is still the highest in the whole. The compressive strength starts to decline when the content of fly ash is more than 35%. When the content of fly ash is 50%, the compressive strength is the lowest and decreases greatly.

Through analysis on compressive strength and flexural strength, it is found that when the content of fly ash is 35%, the compressive strength and flexural strength after soaking are the best. Therefore, when the fly ash content of magnesium phosphate cement is 35%, the micro-aggregate effect and morphological effect reach the maximum among the five contents. The fly ash particles are evenly distributed in the cement slurry, which reduces the porosity, increases the compaction of the cement slurry, and also enhances the fluidity, promotes the expulsion of bubbles in the slurry, and increases the compaction. However, if the fly ash content is too large and the micro-aggregate effect reaches its limit, the excess fly ash will lead to a decrease in strength. When the content of fly ash exceeds 35%, the content of fly ash is too large, and a large amount of unreacted fly ash will lead to a decrease in strength.

3.2. Influence of fly ash content on the immersion durability of magnesium phosphate cement

3.2.1. Influence of fly ash content on the strength retention rate of magnesium phosphate cement

Strength retention ratio is an indicator of water resistance of materials. The water resistance of material directly affects its durability. The compressive strength of magnesium phosphate cement mixed with fly ash was tested by soaking it in static water for 1d* and 7d*, respectively. The compressive strength of specimens soaked in alternative water for 1d** and 7d**, and their strength retention rate was calculated to study the water resistance of MPC mixed with fly ash as shown in Figures 5 and 6.

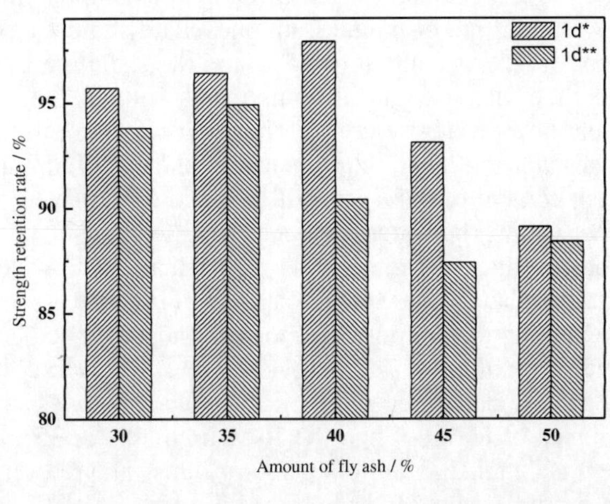

Figure 5. Strength retention rate of fly ash of MPC in soak 1d* and 1d**

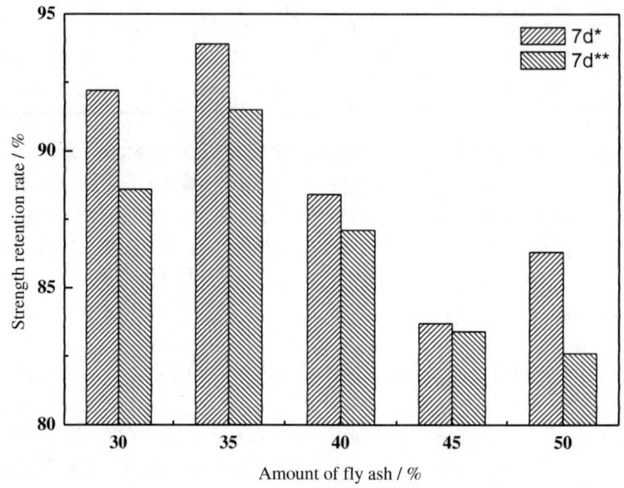

Figure 6. Strength retention rate of fly ash of MPC in soak 7d* and 7d**

As shown in Figure 5, the strength retention rate of fly ash-doped magnesium phosphate cement soaked in water for 1d* and 1d** changes. The strength retention rate of magnesium phosphate cement soaked in static water for 1d* changes with the highest strength retention rate when the fly ash content is 40%, followed by the component with the content of 35%. The trend is to increase first and then decrease. The strength retention rate of magnesium phosphate cement soaked in water for 1d** changes, and the peak value of the strength retention rate does not appear in the component with fly ash content of 40%, but in the component with fly ash content of 35%. The strength retention rate of the components with fly ash content of 50% is the lowest in static water. The strength retention rate of the components with fly ash content of 45% in the immersion water is the lowest compared with the other four groups. Considering the two soaking methods of standing water and changing water comprehensively, the components with a fly ash content of 35% in the five dosages have a better strength retention rate, that is better water resistance. As shown in FIG. 6, the strength retention rate of fly ash-doped magnesium phosphate cement soaked for 7d* and 7d** changes. The overall trend of change is first higher and then lower. For MPC soaked in static water for 7d*, the strength retention rate with the fly ash content of 35% is the highest, followed by the component with the content of 30%. The peak value of strength retention rate of magnesium phosphate cement soaked in water for 7 days also appears in the component of fly ash content of 35%, and the changing trend is first increased and then decreased. At the same time, the strength of components with fly ash content of 50% in static water increased, but the retention rate of strength in changing water was the lowest. When the two soaking methods of standing water and changing water are taken into comprehensive consideration, the components with a fly ash content of 35% are still included in the five contents, and the strength retention rate is better, in other words, the water resistance is better.

Usually, the softening coefficient of the material is higher than 85%, which can be applied in the wet or water-soaked environment, that is, the water resistance of the material is better. The preliminary test found that the strength retention rates of pure magnesium phosphate cement on 1d* and 1d** were 89.4% and 88.1%, respectively, and that on 7d* and 7d** were 81.2% and 79.1%, respectively. The strength retention rate of fly ash - doped magnesium phosphate cement was studied. Therefore, fly ash can improve the water resistance of MPC. When the fly ash content is 35%, the water resistance of magnesium phosphate cement is better than the other four groups. The reason is that when the fly ash content is 35%, it is caused by the activity and physical filling effect of fly ash, which increases the density of magnesium phosphate cement and reduces the porosity. However, when the content exceeds

35%, the micro-aggregate effect of fly ash reaches its limit, and the excess fly ash is in an inert state and does not participate in the reaction to form a dense system. The fly ash is still in a dispersed state, and there are a large number of pores between the particles, resulting in a drop in water resistance. At the same time, because fly ash is composed of oxides of silicon, calcium, aluminum, iron, and other elements, as well as unburned carbon, part of the metal oxides, will participate in the reaction, making the contact part of fly ash and magnesium phosphate cement matrix form a close connection, and consume the phosphate in the MPC slurry. Therefore, adding fly ash can adjust the water resistance of magnesium phosphate cement.

3.2.2. Influence of fly ash content on the quality retention rate of magnesium phosphate cement
Magnesium phosphate cement soaked in water will be eroded. The mass will have varying degrees of loss. According to the retention degree of the soaked magnesia phosphate cement mass, the water resistance of the soaked magnesia phosphate cement is studied directly, which also reflects the water resistance of the soaked magnesia phosphate cement to a certain extent. This paper mainly studies the mass retention of fly ash-doped magnesium phosphate cement soaked for 7d* and 7d** in two soaking methods of static water and changing the water, as shown in Figure 7.

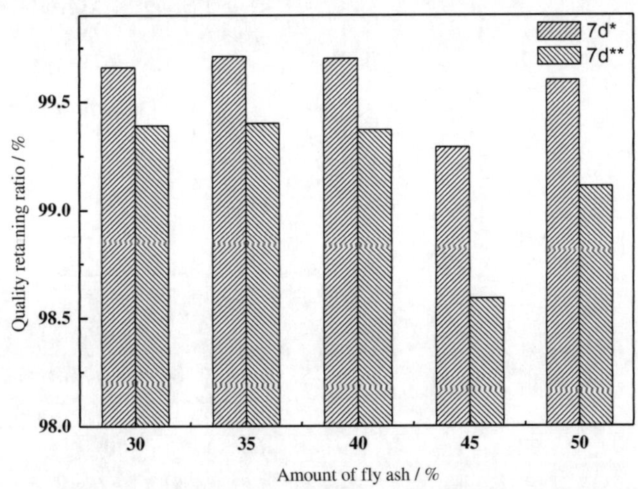

Figure 7. Quality retaining ratio of fly ash of MPC in soak 7d* and 7d**

As shown in Figure 7, represents the change in the mass retention rate of fly ash-doped magnesium phosphate cement soaked for 7d* and 7d**. When the fly ash content is 30%, 35%, and 40%, the mass retention rate is similar. The quality retention rate of the components with fly ash content of 50% is lower than that of the first three groups, while that of the components with fly ash content of 45% is greater than that of the other four groups. This also corresponds indirectly to strength retention. Changes in the mass retention rate of magnesium phosphate cement soaked in water for 7 days ** found that the overall mass retention rate was lower than that soaked in static water for 7 days *, and the peak value appeared in the component with a content of 35%. The overall change rule was similar to that soaked in static water. The mass retention rate of components with fly ash content of 45% drops greatly under the way of changing the water. After comprehensive consideration of the two soaking methods of standing water and changing water, the components with a fly ash content of 35% are still included in the five contents, with good quality retention rate and resistance to water erosion. When fly ash content is 35%, micro-aggregate effect and morphological effect reach the best, which makes the cement slurry dense with fewer structural pores, and to some extent alleviates the dissolution of water on the hydration product of magnesium phosphate cement, and the hydration product is less dissolved. Therefore, when fly ash content is 35%, the quality retention rate of magnesium phosphate cement is higher.

3.3. Mechanism analysis of the influence of fly ash on the water resistance of magnesium phosphate cement

The water resistance of magnesium phosphate cement doped with fly ash is analyzed. On the one hand, the smaller fly ash particles can fill the pores of the cement stone and increase the density of the cement stone itself. On the other hand, since the active SiO_2 of fly ash can react with Mg^{2+} in magnesium phosphate cement to generate $MgSiO_3$.It can not only reduce the concentration of Mg^{2+} in magnesium phosphate cement but also plug the pores in the cement stone to reduce the corrosion of water. Some scholars have also studied that Al_2O_3 in fly ash can also generate $Al(OH)_3$ colloidal substance in magnesium phosphate cement, which plays the same role in cement as $MgSiO_3$. Some scholars have shown that due to the hydrophobicity of fly ash, adding magnesium phosphate cement can reduce the water flow in the cement stone and reduce the concentration of Mg^{2+}.

3.3.1. XRD composition analysis of fly ash - doped magnesium phosphate cement

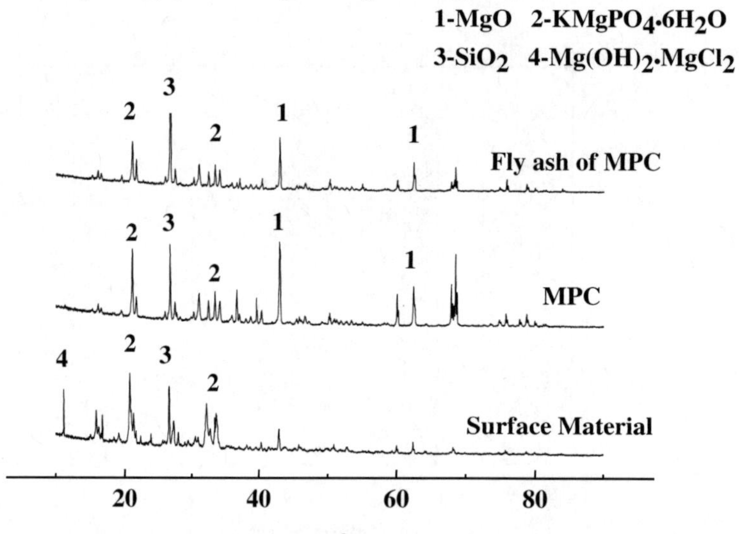

Figure 8. Influence of fly ash on the XRD analysis

Figure 8 is the XRD component analysis of fly ash magnesium phosphate cement. From the XRD curves of pure magnesium phosphate cement powder, the main peaks correspond to unhydrated MgO, SiO2 and KMgPO4•6H2O, respectively. KMgPO4•6H2O is the main hydration product, while SiO2 is carried by fly ash itself. Comparing XRD of fly ash magnesium phosphate cement with pure magnesium phosphate cement powder, it is found that the composition of hydrated material is basically the same, but the main peak is lower than that of pure magnesium phosphate cement. Al(OH)3 is not detected because of the existence of the material itself, but because of the error in the sample preparation process, it is difficult to detect it, which needs further study. There is no MgSiO3 in magnesium phosphate cement mixed with silica fume. The reason is that SiO2 ionizes into SiO32-and needs strong acid environment. The environment after hydration of magnesium phosphate cement is weak alkaline. SiO2 is not allowed to ionize into SiO32-. There is no MgSiO3 in solution. The main peaks of XRD are SiO2, KMgPO4•6H2O and Mg(OH)2•MgCl2. It shows that Mg2+and PO43- ions in the cement body spill out from the cement body after being immersed in water to regenerate bird droppings and other products on the cement surface. There is no Cl- raw material in Mg(OH)2•MgCl2, but the analytical Cl- is introduced by tap water for immersion. According to the generated substances on the surface of cement, researchers found that a part of bird dung stone and unhydrated MgO in magnesium phosphate cement mixed with fly ash would be ionized by water shock to remove Mg2 + and PO43- ions, leaving voids on the surface, which reduced the strength of cement and affected the water resistance of cement.

3.3.2. SEM Micromorphology Analysis of Magnesium Phosphate Cement Mixed with Fly Ash

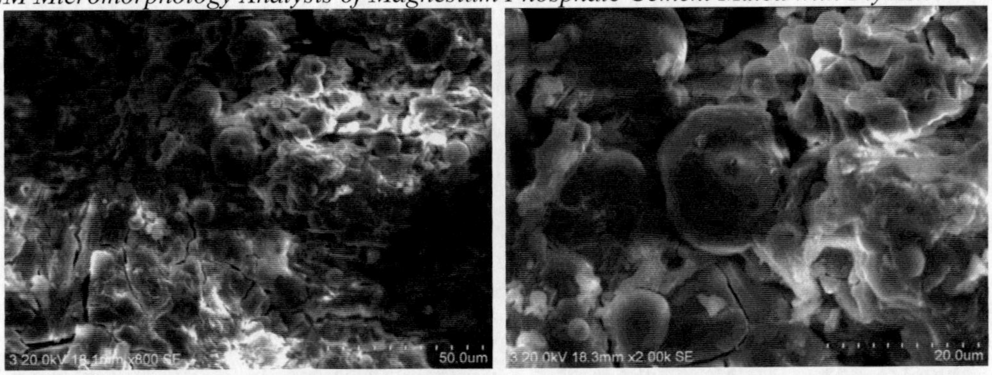

Figure 9. Influence of fly ash on the SEM picture

As shown in Fig. 9, the scanning electron microscopy (SEM) image of the cement mixed with fly ash and magnesium phosphate after 7 days immersion. It can be clearly seen from the Figure that there are many micro-cracks on the surface of the cement paste. These crevices are caused by corrosion of water. At the same time, fly ash particles with different sizes between 5um and 20um can be observed. Fly ash particles are closely connected with magnesium phosphate cement. It shows that fly ash has activity in magnesium phosphate cement. However, the parts closely connected with fly ash are denser and have fewer cracks. Fly ash particles of different sizes can be evenly distributed in the system filled with pore and capillary pore. Improve the pore structure of the structure. In Figure 9, amorphous flocs are formed on the surface of fly ash. This is because the oxides of silicon, calcium, aluminum, iron and other elements in fly ash, as well as the unburned carbon composition, participate in the reaction. The contact part between fly ash and magnesium phosphate cement matrix is closely connected. It also consumes phosphate in MPC paste. The water resistance of magnesium phosphate cement can be improved by using fly ash. Improve the water resistance of magnesium phosphate cement.

4. Conclusion

(1) The mechanical properties of MPC cement can be improved by adding fly ash into MPC cement. The flexural strength and compressive strength of modified MPC after immersion increase with the content of fly ash. They all showed the trend of increasing first and then decreasing. When the content of fly ash is 35%, the compressive strength and flexural strength of modified MPC are the best.

(2) With the increase of fly ash content, the overall trend of strength retention rate and mass retention rate of MPC cement increases first and then decreases, but the magnesium phosphate cement with 50% content slightly rises. When the content of fly ash is 35%, the water resistance is the best, the strength retention rate is more than 95%, and the mass retention rate is more than 99.5%.

(3) The XRD analysis of magnesium phosphate cement mixed with fly ash shows that no new hydration products are formed. The surface substances of the cement paste after soaking are KMgPO4•6H2O and Mg(OH)2•MgCl2. SEM scanning electron microscopy showed that the surface of cement paste had fewer micro-pore, compact structure, better micro-aggregate effect and morphological effect of fly ash. It was evenly distributed in the system, filled with pore and capillary, and improved the water resistance of magnesium phosphate cement.

Acknowledgement

The financial support by Liaoning Innovative Talents Program Project (NO.LR2016060), Liaoning Province's Millions of Talents Project, Shenyang Science and Technology Innovation Platform

Construction Plan (NO.18-006-0-04) and Shenyang Young and Middle-aged Science and Technology Innovation Talents Support Project (RC180136) are gratefully acknowledged.

References

[1] Lin Wei, SUN Wei, Li Zongjin. 2010. Study on the Effects of Fly Ash in Magnesium Phosphate Cement.Journal of Building Materials,13(6), 716-721.

[2] Mao Min,Wang Zhi,Wang Qingzhen,Hu Qianwen,You Chao 2011. Influencing Factors on Water Resistance of Magnesium Phosphate Cement and Its Improving Measures. Fly Ash Comprehensive Utilization (6), 15-18.

[3] Liu Kai, Li Dongxu. 2011. Review of Magnesia-phosphate Cement Based Materials . Materials Reports,25(13), 97-100.

[4] Wang Hongtao, Wu Mingchuan, Tang Chunping. 2005. The effect of Ground Fly Ash on the Properties of Phosphate Cement-based Composites. Fly Ash Comprehensive Utilization (6), 38-40.

[5] Zhang Siyu,Shi Huisheng. 2009. Performances and Applications of Magnesium Phosphate Cement-based Compositesmodified by Fly Ash. Fly Ash Comprehensive Utilization (1), 54-56.

[6] Ding, Z., & Li, Z. 2005. Study of high early strength cement based on fly ash, magnesia and phosphate.Materials Technology,20(3), 136-141.

[7] Zhou Xuyang, Yang Jianming.2013. Effect of Fly Ash on Properties of Magnesium Potassium Phosphate Cement Paste.Journal of Jiangsu University (Natural Science Edition), 2013, 34 (6):730-735.

[8] Sarkar A K.1990. Phosphate Cement-based Fast-setting Binders [J]. Ceram Bull,69(2): 234-238.

[9] Luo Yali.2010 Experimental Study on Magnesium Phosphate Cement With High Early Strength and its Applications(Shanghai: Shanghai Jiao Tong University).

[10] Mao Min, Wang Zhi, Wang Qingzhen.2011 Influencing Factors on Water Resistance of Magnesium Phosphate Cement and Its Improving Measures. Fly Ash Comprehensive Utilization, 15-18.

[11] Nelson N O, Mikkelsen R L , Hesterberg D L., 2003 Struvite Precipitation in Anaerobic Swine Lagoon Liquid: Effect of pH and Mg:P Ratio and Determination of Rate Constant. Bioresource Technology, 89(3) :229 – 236.

[12] Y Quanbing, Z. Beirony, Zh. Shuqing etal., 2000, Poroperties and Application of Magnesia-phosphate Cement Mortar for Rapid Repair of Concrete. C.C.R 30(11): 1807-1813.

[13] Li Dongxu, Li Pengxiao, Feng Chunhua. Research on Water Resistance of Magnesium Phosphate Cement. Journal of Building Materials, 2009(9):505-508 525.

[14] D.A.Hall, R.Stevens, B.El-Jazairi. 2001. The Effect of Retarders on the Microstructure and Mech-apical Properties of Magnesia-phosphate Cements Mortar. C.C.R, 31(3):389-396.

Effect of particle size distribution of slag on the strength and pore structure of low-temperature concrete

Yao Li[1], Na Teng[2], Runqing Liu[1*] and Yunpeng Cui[3]

[1]School of material science and engineering, Shenyang Ligong University, Nanping Road No.6, Shenyang, China, 110159
[2]School of materials science and engineering, Shenyang JianZhu University, New Hunnan District, Shenyang, China, 110168
[3] Faculty of Infrastructure Engineering Dalian University of Technology, Dalian, China, 116024

*Corresponding author and e-mail: Liu Runqing, 13940195514@163.com

Abstract. In order to study the effect of particle size distribution of slag on strength and pore structure of the concrete at low temperature, the particle sizes of slag were divided into four grades, namely, 5~10 μm, 10~15 μm, 15~20 μm and 20~26 μm. The influence of particle size distribution of slag on the compressive strength and pore structure of low temperature concrete cured for 3 days, 7 days and 14 days were analyzed. The results show that different particle size distribution of slag leads to different pore structure and strength of concrete. When the slag with the size is of 10～15μm, the concrete obtains the best performance, regardless of curing temperature. In addition, the decreasing granularity of slag improves the pore structure and increases the density of concrete.

1. Introduction

In the cold region of China, the winter lasts for a long time, which is unfavourable to concrete construction and curing. In general, there are ways to solve this problem, such as the curing equipments was used to improve its insulation performance, or mixed with admixtures as silica fume, slag powder, fly ash and so on.

Ground granulate blast furnace slag is a by-product from the iron and steel manufacturing industry. Its main ingredients are SiO2, CaO and Al2O3. Generally, volume fraction of the three ingredients can reach more than 90%[3-4]. Slag is a normal mineral admixture, due to the higher activity as compared to fly ash.

2. Experimental materials and methods

The cement used in this test is P O 42.5 grade ordinary Portland cement produced by Anshan Jidong cement co., LTD. The properties of the Portland cement are shown in Table 1. Granulated blast furnace slag powder was produced by Anshan iron and steel co., LTD. The chemical composition is shown in table 2.The sand is medium sand with sand rate of 0.3. Polycarboxylic acid superplasticizer was mixed with a content of 1.0% of the cementing material, and the performance index met the requirements of GB 8076-2008. Sodium nitrite was selected as antifreeze agent, and the content was 3.0% of the cementing material.

Table 1. Properties of the Portland cement

label	stability	Setting time/min		Compressive strength/MPa		Flexural strength/MPa	
		Initial set	Final set	3 d	28 d	3 d	28 d
P. O 42.5	qualified	3:18	4:24	23.6	52.3	5.7	9.6

Table 2. Chemical composition of slag /%

Composition	CaO	SiO2	Fe2O3	Al2O3	MgO
Content	32.56	42.56	3.65	14.70	7.45

Based on our preliminary test, when the slag content is 20%, the concrete strength reaches the optimal value. The slag dosage in this work was thus fixed as 20%, and the mix design is shown in table 3.

Table 3. Mix design/ (kg/m^3)

Water	Cement	Slag	Sand	Crushed stone	Water-reducing agent	Antifreeze
128	328.2	82.05	585.52	1244.23	4.1	12.3

Slag was screened by standard sieve with different mesh size, and the mesh size corresponding to pore size was shown in Table 4.The particle size of the slag passed through screen sieves to obtain particle sizes of 20～26μm, 15～20 μm, 10～15μm and 5～10μm.

Table 4. Screen mesh number and corresponding aperture

Mesh	The aperture/μm
2300	5
1800	10
1000	15
900	20
600	26

The concrete specimen (100 mm * 100 mm * 100 mm) were cured at 10 °C, -10 ～10 °C and -10 °C, and was taken out for compressive strength test after cured to 3 d, 7 d and 14 d curing. In order to characterize the meso-pore structure of concrete, samples after the compressive strength were crushed and analyzed by mercury intrusion method.

3. Experimental results and analysis

3.1. Influence of slag size and curing temperature on compressive strength of concrete

The slag with particle size of 5 ~ 10μm, 10 ~ 15μm, 15 ~ 20μm and 20 ~ 26μm were added to the concrete respectively, and its compressive strength was tested after curing for 3d, 7d and 14d. The results are shown in Fig. 1.

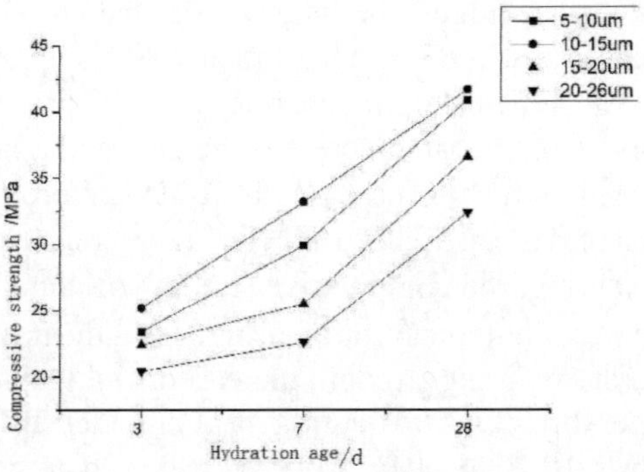

(a) Curing temperature is 10 °C

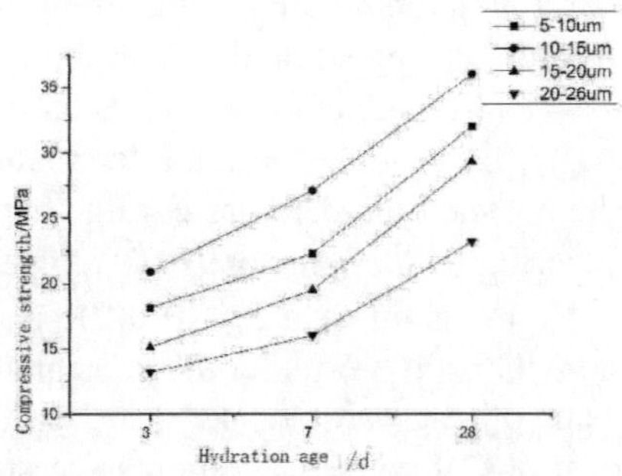

(b) Curing temperature is -10 °C～10 °C

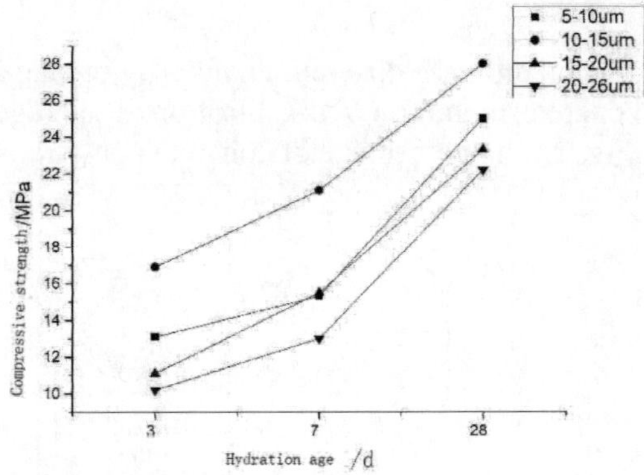

(c) Curing temperature is -10 °C

Figure 1. Compressive strength of concrete with different slag particle size distribution

As seen in Fig. 1,at 10 °C temperature curing temperature slag granularity for 10 ~ 15 μm of 3 d, 7 d, 28 d compressive strength is the largest, the biggest compressive strength stronger than 5 ~ 10μm large 1.9%, the biggest compressive strength stronger than 15 ~ 20μm large 13.93%, stronger than 20 ~ 26μm large 28.7%.The compressive strength increases with the curing age. In -10 ～ 10 °C temperature curing, the slag particle size for 10～15μm of 3 d, 7 d, 28 d compressive strength is the largest, the biggest compressive strength stronger than 5 ～10 μm large 12.5%, stronger than 15～20μm large 22.87%, stronger than 20 ～26μm large 55.17%.The compressive strength increases with the curing age. In - 10 °C temperature curing environment, slag granularity for 10～15 μm of 3 d, 7 d, 28 d compressive strength is the largest, the biggest compressive strength stronger than 5～10 μm large 12%, stronger than 15～20 μm large 20.17%, more than 20～26μm large 26.13%.T The compressive strength increases with the curing age, because of the filling effect of slag in concrete, the fine slag powder particles are filled in the void between clinker particles, which increases the bulk density of powder and enhances the density of cement slurry, thus leading to the increase of cement strength [7-9]. So the size of the slag will affect the strength of concrete, therefore 10～15μm slag is the most suitable for concrete void.

When slag in concrete size for 5 ～10μm, curing temperature at 10 °C temperature, the compressive strength of concrete is the largest,, the largest stronger than in -10～10 °C under variable temperature curing strength increased by 27.8%, stronger than - 10 °C under the constant temperature curing strength increased by 55.17%. When slag in concrete size for 10～15 μm, curing temperature at

10 °C temperature, the compressive strength of concrete is the larges, the largest more than in -10 ～ 10 °C under variable temperature curing strength increased by 15.83%,more than - 10 °C under the constant temperature curing strength increased by 48.93%. When slag in concrete size for 15～20μm, curing temperature at 10 °C temperature, the compressive strength of concrete is the larges, the largest more than in -10～10 °C under variable temperature curing strength increased by 24.9%, more than - 10 °C under the constant temperature curing strength increased by 57.08%.When slag in concrete size of 20～26μm, curing temperature at 10 °C temperature curing, the compressive strength of concrete is the larges, the largest more than in -10～10 °C under variable temperature curing strength increased by 39.65%, more than - 10 °C under the constant temperature curing strength increased by 45.95%.

In four groups of slag particle size distribution range, three kinds of curing system under different curing ages relations of concrete compressive strength is 10 °C > -10～10 °C > - 10 °C.The growth rate of concrete strength increases with age, and the lower the temperature is, the higher the growth rate of concrete strength is.

3.2. Influence of slag particle size on concrete pore structure

In curing temperature 10 °C, the mixed with different granularity of slag 10 ~ 15μm, 15 ~ 20μm of the total pore volume, total of concrete hydration 7 d control area, average pore diameter and porosity were measured, as shown in Fig. 2. The percentage of concrete aperture is shown in Fig.3

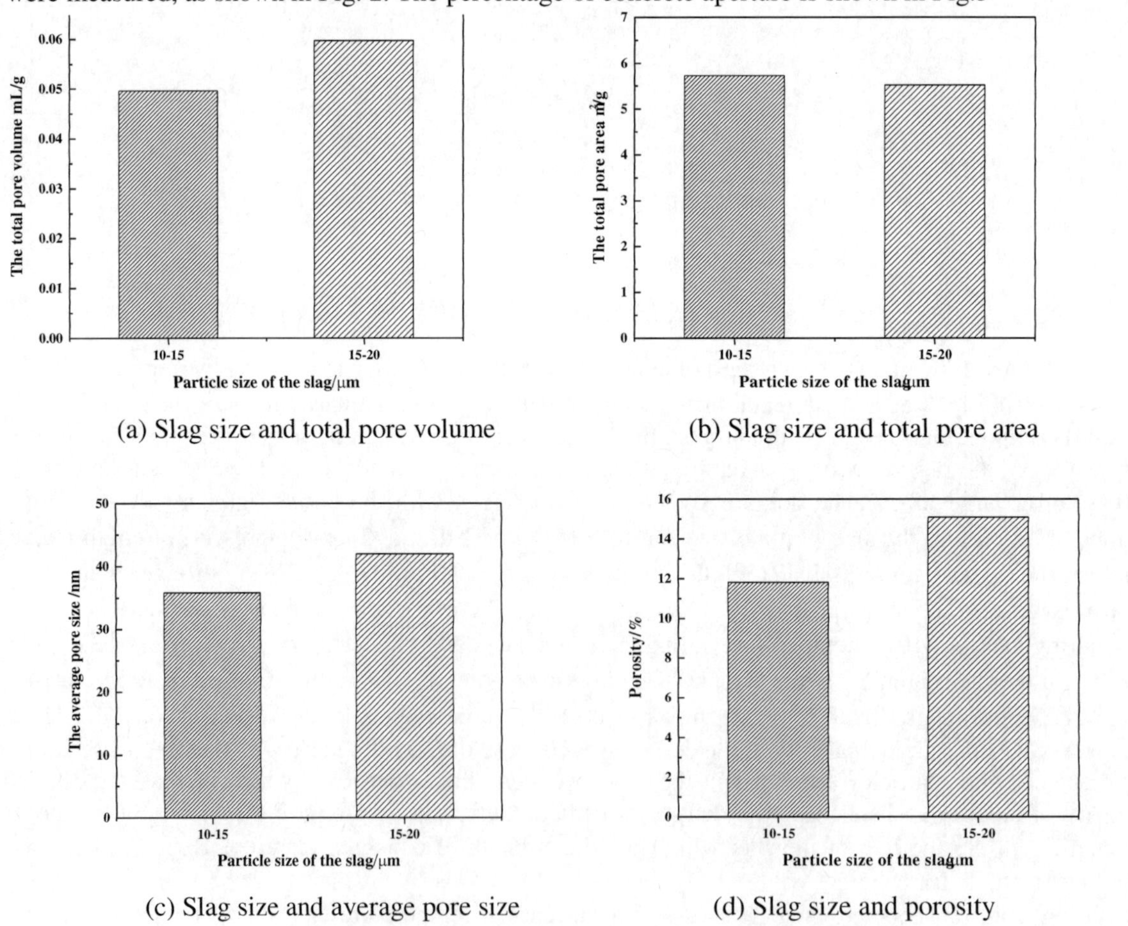

(a) Slag size and total pore volume (b) Slag size and total pore area

(c) Slag size and average pore size (d) Slag size and porosity

Figure 2. characteristic parameters of slag size and void

According to Fig. 2 (a), the total pore volume of concrete increases with the increase of slag particle size; It can be seen from Fig. 2 (b) that the total pore area of concrete decreases with the increase of slag size. As shown in Fig. 2 (c), the average pore size of concrete increases with the increase of slag size; According to Fig. 2 (d), the porosity of concrete increases with the increase of slag size.

So, slag, the smaller the particle size, density of slag can fill into the concrete, slag increasing participation in secondary hydration reaction, generate more hydration products fill in the porosity of the concrete hydration products, break up a big hole in the slurry, constantly into numerous harmless puncture, slurry structure to reduce the gap, the average pore diameter reduced gradually, raise the compressive strength of samples. That is, the smaller the slag size, the greater the strength[10-11].

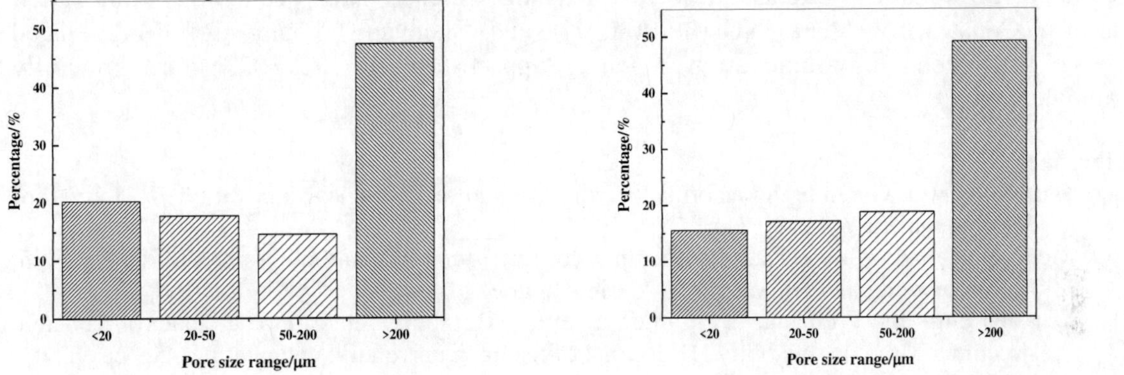

(a) Percentage of concrete aperture with particle size of 10-15μm

(b) Percentage of concrete aperture with particle size of 15-20 μm

Figure 3. Slag size and pore size percentage

Fig. 3 (a) shows the percentage of concrete pore sizes at different levels within the particle size distribution of slag within 10 ~ 15μm, Among them, those less than 20nm accounted for 20.42%, 20-50nm accounted for17.76%, 50-200nm accounted for 14.56%, and more then 200nm accounted for 47.25%. Fig. (b) shows the percentage of pore diameter of concrete at all levels within the range of slag size distribution of 15-20μm, those less than 20nm accounted for 15.67%, 20-50nm accounted for17.21%, 50-200nm accounted for 18.84%, and more then 200nm accounted for 49.28%.Therefore, the particle size of mineral admixtures in concrete is different from each other. It can be seen that when the particle size of slag is distributed at 10 ~ 15μm, The content of less harmful pores under 20nm is greatly increased, the content of harmful holes and multi-harmful holes larger than 50nm was relatively reduced, because the larger particle size of concrete mineral admixtures is, the more favorable it is for the formation of multi-damage pore size in concrete, the percentage of large holes formed in concrete is also high, It is also dangerous to concrete. Therefore, the use of smaller slag particle size, so that the concrete pore structure has been significantly improved, slurry structure more dense, lower porosity, strength has been improved.

4. Conclusions

(1) The compressive strength of concrete varies in the order of 10~15μm>5~10 μm>15~20μm>20~26μm for all the curing ages. Consequently, there is an optimal value of particle size distribution (10 ~15μm) at which the highest compressive strength of concrete can be obtained.

(2) The higher the curing temperature, the lager the compressive strength of concrete is. The compressive strength of concrete varies in the order of 10 °C > - 10 ~ 10°C > - 10°C for all the particle size distribution of slag.

(3) Total pore volume, average pore diameter and porosity increases with increase in the particle size of slag for all the curing temperature, when the total pore area decreases.

(4) The features of the pore structure of concrete are few big harmful pores, more harmless pores, when the particle size of slag was 10-15μm which is beneficial to improve concrete strength.

IOP Conf. Series: Materials Science and Engineering **587** (2019) 012008 doi:10.1088/1757-899X/587/1/012008

Acknowledgement
About the author: Li Yao, female, born in 1983, doctor, teacher, Shenyang, Liaoning Province. Address: School of Science Materials and Engineering, Shenyang Ligong University, Nanping Road No.6, New Hunnan District, Shenyang, China; E-mail:49728418@qq.com.

The financial support by the Natural Science Foundation of Liaoning Province, China, (20180550329), Liaoning Innovative Talents Program Project (NO.LR2016060), Liaoning Province's Millions of Talents Project, Shenyang Science and Technology Innovation Platform Construction Plan (NO.18-006-0-04) and Shenyang Young and Middle-aged Science and Technology Innovation Talents Support Project (RC180136) are gratefully acknowledged.

References
[1] Wu Zhong wei. Green high-performance concrete and technological innovation [J]. Journal of building materials, 1998,1 (1):1-7.
[2] Meng Qingchao. study on factors affecting concrete durability and pore structure [D]. Harbin. Harbin Institute of Technology, 2006. (in Chinese)
[3] Wu Hairong, Jin Weiliang, Lv Qingfang et al. Durability environment zone of concrete structures based on reliability [J]. Journal of zhejiang university: engineering science edition. 2012,46 (003): 416-423.
[4] Liu run qing. Research on the frost critical strength of low-temperature concrete under the influence of multiple factors [D]. Dalian. Dalian university of technology, 2011.
[5] Liu runqing, sun si hui, Yang yuan quan. Influence of fly ash particle size distribution on hydration performance of Portland cement [J]. China powder technology,2017,23(05):83-86.
[6] Jiang Yonghui, Yan Chunxia. grey system study on influence of fly ash particle distribution on cement strength. Acta silicate sinica, 1998, 26(4): 194-198. (in Chinese)
[7] ZHANG G,LI G. Effects of mineral admixtures and additional gypsumon the expansion performance of sulphoaluminate expansive agent atsimulation of mass concrete environment[J]. Construction and Building Materials,2016,113: 970-978.
[8] Bertil Persson. Internal frost resistance and salt frost scaling of self-compacting concrete[J]. Cement and concrete Research, 2003(4):373-379.
[9] Zhang Y M, Napier-Munn T J. Effects of particle size distribution, surface area chemical composition on Portland cement strength[J]. Powder Technology, 1995, 83(3): 245-252.
[10] Wang Aiqin, Zhang Chengzhi, Zhang Ning sheng. The theoretic analysis of the influence of the particle size distribution of cement system on the property of cement[J]. Cement and Concrete Resesrch, 1999: 121-126.
[11] Celik I b. The effects of particle size distribution and surface area upon cement strength development[J]. Powder Technology, 2009, 188(3): 27.

IOP Conf. Series: Materials Science and Engineering **587** (2019) 012009 doi:10.1088/1757-899X/587/1/012009

Experimental study on material performance of separation grouting for prevention of rock burst

Zhangliang Chen*, Junwei Shi, Binbin Zheng

School of Management Science And Engineering, Shandong Institute of Business And Technology, Yantai, China

*Corresponding author and e-mail: Zhangliang Chen, chenzhl_yt@126.com

Abstract. The purpose of this study is to obtain the optimal proportion of the full tailings cement filling, reduce the filling cost of the mine and ensure the stability of the empty area. Taking the whole tailings cemented backing block as the research object, the uniaxial compressive strength of the test block as the index, the filling body is investigated comprehensively. The influencing factors of strength were studied by orthogonal experiment, and the sensitivity and significance of the influence factors of the filling body strength were analyzed by using the range and variance analysis methods. The results show that, when cement content is increased, paste setting time is shorten and paste strength is enhanced; proper quantity of fly ash can improve the liquidity of paste, while degree of segregation and bleeding rate of paste are increased with the increase of fly ash; When cement ratio is 20% and the rate of fly ash: coal gangue is 1: 4 and the concentration of the paste is 75%, the pumpability and strength of paste is the best.

1. Introduction

Impact rock pressure refers to the dynamic phenomenon of sudden strong damage caused by the elastic deformation energy released by the rock mass around the mine roadway or working face. The occurrence of impact ground pressure is often accompanied by the throwing of coal rock mass. Huge sounds and impacts such as air waves [1], impact ground pressure has the characteristics of "strong disturbance" and "strong aging", which is a typical power disaster in coal mining. At present, fly ash tends to be cement clinker in construction engineering. Other auxiliary active materials are used together to produce cement and formulated concrete, but there are few applications in coal mining cementation filling method. Therefore, the author proposes green cement filling material based on fly ash-gangue. The research uses a fly ash-cement-desulfurization gypsum composite gelling system with fly ash as the main component, the original fly ash and the crushed coal gangue as the aggregate, and the orthogonal test design method is used for material mix ratio. Optimized design, strive to maximize the use of fly ash, desulfurization gypsum and other industrial waste to replace cement for cementation filling, for the preparation and promotion of high-quality low-cost cementation filling materials to make a useful exploration.

2. Experimental study on optimization of proportion of vermiculite cement filling materials

At present, the most effective way to control the surface subsidence of coal mining is filling and overburden grouting. In terms of filling and mining, the investment cost of filling pipeline system, filling equipment and filling materials is too high [2]. Therefore, the overburden grouting technology is getting more and more attention, especially at the current stage of China's coal industry. Under the grim situation of overall profit decline. The separation grouting technology also has two problems to be solved: First, due to the influence of mining, the overburden movement is a dynamic process, and the height and width of the separation layer are constantly changing with the mining. Therefore, The location and timing of layer grouting are difficult to grasp. Once the mining is separated, the grouting work cannot be continued. This reduces the total grouting and does not achieve the desired effect in controlling surface deformation [3]. Second, the grouting material is generally made of fly ash. The mortar mixed with water, and the cementing ability of fly ash is weak. Once the mortar loses water in the layer, the consolidated fly ash is difficult to ensure sufficient strength to limit the movement of the overburden and control the deformation of the surface. Third, the water-cement ratio of the leaved grouting material is generally 2:1, and the water content is about 70%. After the fly ash in the separated zone is dehydrated, so much water will enter the stop face through the crack zone. The working surface environment has an adverse effect and even the danger of flooding.

2.1. Project overview
Study area is selected Hua Feng mine 1410 working face, the mining as deep as 1140m, coal seam thickness of the overlying thick conglomerate of 400 ~ 800 meters. Under the influence of mining activities, the thick conglomerate layer (about 800m)fracture movement makes the frequency and intensity of rock bursts occurred in rising, which threaten to the safety of the miners. The working face is located in the mid-eastern of the field, and coal group second segment before the first mining area of a -1100m level, the high limit and the low limit were -840m and -920m respectively, and working face from west to east on mining, for 1409 fully-mechanized caving goaf, next to un-mined 1411 working face. The main mining coal seam in 4# in this area for the structure is relatively simple thick coal seam, thickness in 5.6 ~ 6.9 m, 6.2 m on average [4]. The main mining coal seam in 4# has serious impact tendency, which has been in more than one face repeatedly impact ground pressure. Thick conglomerate layer of covering above the mining field (about 800m) integrity is stronger. The movement of thick conglomerate and multilayer cantilever beam, which will produce the markings in the surface subsidence, and has enormous influence on farmland, riverbed and buildings.

2.2. Material composition of the model
Similar materials mainly include two aspects of raw materials, padding (or the aggregate) and cement. Normally, Padding uses river sand, mica powder, talc, etc., and cement includes gypsum, paraffin, calcium carbonate, cement, etc. For simulation of the overlying stratum movement, the experiment of similar material cement is plaster, meanwhile, adding calcium carbonate to increase its strength. The padding is river sand, and scatters mica powder between each layer that play a part in layering [5]. The difference with mock object and ratio, can be realized by different ratio of similar materials [6].

The choice of filling materials must follow two principles: first, the source is extensive; second, the cost is lower. In this test, the fly ash of the power plant of Xinzhuang Coal Mine was selected as the fine aggregate, the specific surface area was 750 m2/kg, the residual amount of 0.045mm square hole sieve (the original gray scale) was 58%, and the standard consistency water demand ratio was 135%. Desulfurization gypsum and 325 ordinary Portland cement clinker are mixed as a composite cementitious material in a ratio of 2:1; the coarse aggregate is a secondary crushed coal gangue in Xinzhuang Coal Mine, the water absorption rate is 6%, and the crushing index is 13.2%. The determination of the particle size grading of coal gangue was carried out by randomly sampling the vermiculite (particle size ≤ 25 mm) before the secondary crushing and mixing in the Xinzhuang Coal Mine. The test selected 10 square holes with a hole diameter of 25mm, 20mm, 16mm, 13.2mm, 9.5mm, 5mm, 1mm, 0.3mm, 0.1mm, 0.075mm, etc., to screen the coal gangue, and record each time. Screening allowance, and calculate the percentage of sieves with different particle size.

IOP Conf. Series: Materials Science and Engineering **587** (2019) 012009 doi:10.1088/1757-899X/587/1/012009

2.3. Similar models production and excavation

The experiment in Huafeng coal mine, xinwen mining bureau 1410 working face.conditions of enclosing,rock for the prototype, making a two-dimensional model. Model constituted by the coal seam,floor,and overlying strata. This experiment adopts similitude simulation test-bed systems, consisted of frame system, loading system and test systems. Framework specification for 2m*0.2m*2m, based on similar criteria,and geosim is 1:200.According to the overburden failure theory, model height should be greater than half of the mining areas. According to the analysis of the geological data, model runs up to 17 layers, and height of 1.46 m, and scattering mica powder between each layer that play a part in layering.Top model adopts hydraulic system through the joist steel of the uniform load(Q = 15.22 MP,r=0.025MN/m^3). Coal seam excavation along the tunneling of the coal floor, every time moves the 5cm, equivalent to the actual propulsion 10m at a time. According to similar principles and total of excavation of 15cm a day, equivalent to the actual mining 30m.Model ends retain the 40m boundary pillar in order to eliminate the influence of boundary conditions.

3. Grouting method for sac-type separated layer grouting material for controlling surface subsidence

3.1. Filling material grouting process

(1) Preparation for construction: the site is leveled, and the positions of several grouting holes are staked, and the spacing between adjacent grouting holes is 10~30m;

(2) drilling: drilling the grouting hole to a design depth, the opening diameter of the grouting hole is 240~300mm, and the inclination of the grouting hole is less than 1%;

(3) Grouting pipe and bag making: the length of the grouting pipe is larger than the designed depth of the hole 400~600mm; the upper end of the grouting pipe is provided with a grouting pipe elbow, and the grouting pipe elbow is connected with the high-pressure rubber a plurality of grouting holes in the shape of a plum blossom are arranged at a distance of 300 to 400 mm from the lower end of the grouting pipe, and a rubber valve plug is used at the lower end of the grouting pipe to seal the bottom; the diameter of the bag is 500~ 600mm, the length is 600~700mm more than the drilling design depth;

(4) the grouting tube and the pouch are tied, the lower hole: the pouch is first sleeved from the bottom to the outside of the grouting tube, the bottom of the pouch is fastened with a thin wire, and then the pouch is The bottom end of the bag is folded back 250~300mm, and then fastened with fine wire 19; from the bottom end of the bag, every 0.8~1.2m, the bag is tied with the rubber band or 18 wire. On the slurry pipe, at the uppermost end of the bag, the rubber collar is fastened to ensure that the plastic collar can slide relative to the grouting pipe, and the upper end of the grouting pipe protrudes above the ground by more than 500 mm; After the grouting hole is cleaned, the grouting pipe to which the pouch is attached is slowly placed into the grouting hole, and when the pouch reaches the design height, the lowering is stopped;

(5) Mixing and grouting: the cement, fly ash, admixture and mine drainage slurry stirred in the mixing tank enters the high pressure grouting pump, and the ground is pushed by the high pressure grouting pump a pumping pipe, the high pressure rubber hose, the grouting pipe elbow and the grouting pipe inject slurry into the bladder, during which time a pressure gauge of the high pressure grouting pump is used to ensure The grouting pressure is 0.4~0.6 MPa, and the flow rate of the grouting is controlled by the electromagnetic flowmeter on the ground pumping pipeline to be 50-65L/min; the bottom-up grouting is to be injected into the upper section of the pouch. After the slurry is designed to the amount of pulp, the next section can be grouted;

(6) Pulling the slurry pipe: after the bag grouting amount reaches the design requirement, and when the grouting amount and the grouting pressure are substantially the same as the design requirements, the bag bag has fully expanded to expand the bag, The grouting pipe is slowly pulled out, and the speed of pulling the grouting pipe is controlled at 0.5~0. 8m/min, and the grouting pressure is maintained at 0.4~0.6 MPa; after the grouting pipe is completely pulled out, Backfill the upper hole with sand and gravel, and compact the ground vibration;

(7) Flush the grouting tube and move to the next hole for the next cycle.

3.2. Advantages of grouting materials

(1) The time of grouting in the layer can be flexibly controlled. In the early stage of coal seam mining, the pocket-like grouting reinforcement can be carried out under the key layer which is easy to produce, which is not affected by the dynamic change of the separation space in the mining. .

(2) Different from the traditional separation grouting process, the bag-type separation layer grouting method of the present invention is actually a separation layer reinforcement process, which performs lateral extrusion and longitudinal reinforcement on the separation layer under the key layer. The soft rock layer in the separation zone has certain bending, shearing and compressive strength, and the traditional passive post-separation grouting is the active separation of the layer, which fundamentally limits the development of the separation layer and ensures that the key layer is not Will break, to achieve the purpose of controlling the surface subsidence.

(3) Through the isolation of the bladder, the slurry will not leak, spread to the working surface through the crack, and pollute the environment.

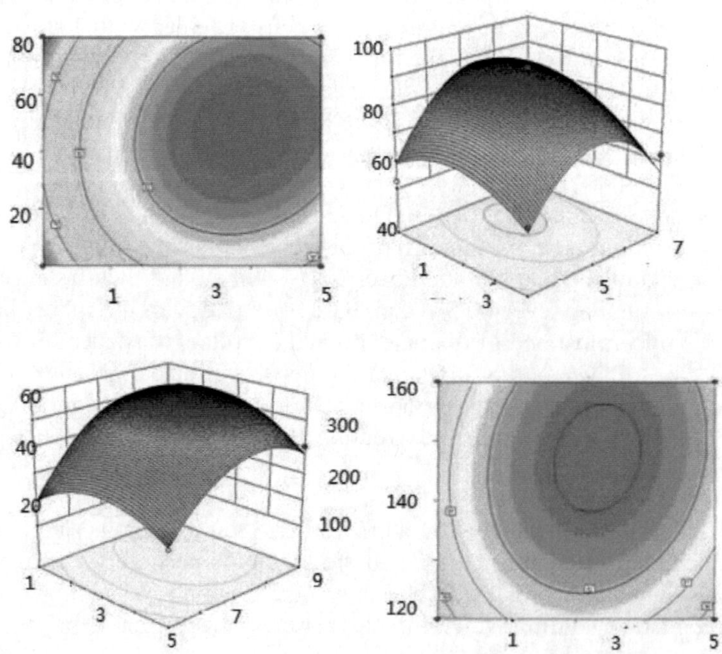

Figure 1. Response surfaces of A and B and C to any two significant influential factors

3.3. Grouting material performance evaluation

After the grouting is completed, part of the grouting material is taken for laboratory testing, and the setting time, slumping degree, stratification degree and bleeding rate are measured and recorded. Three sets of experiments are carried out and three tests are carried out in each set of tests. The size of the block and the test block is (150.0 × 150.0 × 150.0) mm3. After demoulding, it is cured in a constant temperature curing box with a curing temperature of 22 ° C and a humidity of 85%. After 28 days of maintenance, after the specified age, the large test piece is demolded, and the large sample module is cored and polished by using the test sample processing equipment (AHM-200) and cutting machine (DQ-4). Since the rock mechanics test specification stipulates that the standard rock sample is a cylindrical shape of 50mm*100mm, this time adopts a 50mm drill bit (inner diameter 50mm, that is, the core diameter is 50mm) to take the core, and finally the YAW-400 type pressure testing machine is used to measure the single. The axial compressive strength, the average of each set of experimental results is shown in Table 1.

Table 1 Grouting material performance evaluation form

Index / Number	Setting time/h	Slump/cm	Stratification/mm	Bleeding rate/%	Compressive strength/MPa	
					8h	28d
S1	3.1	23.7	15.0	1.9	0.67	3.84
S2	2.8	22.3	11.0	1.5	0.85	4.27
S3	2.5	21.6	13.0	1.4	0.84	5.1

4. Conclusions

This paper provides a bag-type separation layer grouting material for controlling surface subsidence, which is prepared from the following components by weight ratio: cement, fly ash, water and admixture, wherein the water-cement ratio is 0.7-0.8. Fly ash: cement is 0.8~0.9, and admixtures account for 3%~5% of the total. The bag type separation layer grouting material of the invention has good pumpability, low bleeding rate and stratification degree, short coagulation time, high strength and stability, less grouting amount, and total single grouting It is about 13m3 and the material cost is low. The invention also provides a preparation method and a grouting method for the above-mentioned bag type separation layer grouting material, the preparation method is simple in operation and low in cost; the grouting method can be flexibly controlled by the layer grouting time, and fundamentally Limit the development of the separation layer, ensure that the key layer will not break, and achieve the purpose of controlling the surface subsidence. Through the isolation of the capsular bag, the slurry will not leak and pollute the environment.

Acknowledgement

This research was financially supported by Study on Physical and Mechanical Properties of High Thickened Tailings during Upstream Dam Filling of National Natural Science Foundation Project(Grant NO.51804178), Teaching Reform Project of Shandong Business School: A Study on Hybrid Teaching Model and Evaluation System Based on Application-oriented Talent Cultivation (Grant NO. 1168G201701) and Shandong Natural Science Foundation: Research on enterprise informatization introduction mode and enterprise value increment oriented to supply and demand network(Grant NO. ZR2012GM003).

References

[1] F K CHENG, J P YAN, T Y FAN, et al. Research on optimizing operation parameters for leaching chlorophyll-a of landscape water based on orthogonal experiments[J]. Ecological Science, 2014, 33(6): 1085-1090. (In Chinese)

IOP Conf. Series: Materials Science and Engineering **587** (2019) 012009 doi:10.1088/1757-899X/587/1/012009

[2] L M DOU.J L XU,C P LU. The abscission grouting control rock burst risk mechanism's discussion[J]. Journal of China University of Mining and Technology,2004,33(2):145 – 149. (In Chinese)

[3] JAMES C N, CHENG Wenchieh. G rout efficiency of lifting structure in soft clay [J] . ASCE, 2009.

[4] J W SHI, Z J WEI , Q L LIU. Research on Proportion Optimizing of Paste Filling Material Based on Orthogonal Experiment in Coal Mine [J]. Journal of China University of Mining and Technology, 2011, 21(6): 112-115. (In Chinese)

[5] Palomo A, Fernandez-Jimenez A, Criado M. Microstructure development of alkali- activated fly ash cement: a descriptive model [J]. Cement and Concrete Research, 2005(35):1204-1209.

[6] Huynh L, Beattie D A, Fornasiero D, etal. Effect of Poly phosphate and Naphthalene Sulfonate form aldehyde Condensate on the Rheological Properties of Dewatered Tailings and Cemented Paste backfill[J]. Minerals Engineering, 2006, 9(1):28-36.

IOP Conf. Series: Materials Science and Engineering **587** (2019) 012010 doi:10.1088/1757-899X/587/1/012010

Adsorption of Cr(VI) from aqueous solution by crosslinked cationic konjac glucomannan with different degree of substitution

Pengyuan Wang, Ruolin Wang, Jianye Liu, Chunmei Niu*

School of Material Science and Engineering, Hebei University of Science and Technology, Shijiazhuang 050018, China

*Corresponding author and e-mail: sjzncm@163.com

Abstract. Chromium has numerous industrial applications, such as metallurgy, processing of leather, producing chromic acid, acting as catalyst and electroplating. These activities cause Cr(VI) contamination in air, soil and water resources. Developing efficient and cost- effective adsorbents for removing Cr(VI) is of utmost importance. Crosslinked cationic konjac glucomannan (CCKGM) with different degrees of substitution (DS) 0.226, 0.391 and 0.456, were prepared by reaction of 3-chloro-2-hydroxypropyl-trimethyl ammonium chloride, konjac glucomannan (KGM) and epichlorohydrin and used to adsorb Cr(VI) ions from the aqueous solutions. Then the adsorption behaviour of CCKGM to Cr (VI) was explored. The adsorption effect of CCKGM on Cr(VI) is affected by pH, Cr(VI) initial concentration, adsorbent dose and temperature. With the increase of DS, adsorption capacity increased. The adsorption capacity was 31.52 mg/g, 58.82 mg/g and 62.59 mg/g corresponding to DS 0.226, 0.391 and 0.456, respectively. Adsorption can get equilibrium within several minutes and follows Langmuir isothermal adsorption. The process is exothermal and low temperature is favourable for adsorption. CCKGM has potential to be used as adsorbents for Cr (VI).

1. Introduction

Environmental pollution caused by toxic heavy metal in industrial effluents is one of the most pressing problems in the world. Water pollution due to chromium compounds remains a serious environmental problem. Chromium ions are often found in the environment as a result of their industrial use, such as electroplating, metal finishing, leather tanning, pigments etc. Chromium(VI) compounds are known to be 1000 times more toxic than chromium(III) and cause allergy,skin ulveration and damage to the liver, kidneys and blood cells.

Strict legislation on the discharge of these toxic products makes it necessary to develop various efficient technologies for the removal of pollutants from wastewater [1]. Different technologies and processes are currently used. Biological treatments [2], membrane processes [3], advanced oxidation processes [4], chemical and electrochemical techniques [5, 6] and adsorption procedures [7-9] are the most widely used for removing metals from industrial effluents. Among all the treatment propositions, adsorption is one of the most popular methods. It is now recognized as an effective, efficient and economic method for water decontamination application and separation analysis. Recently, numerous approaches have been studied for the development of cheaper and more effective adsorbents containing natural polymers. Polysaccharides which are abundant, renewable and biodegradable resources are able to associate by physical and chemical interaction [10,11]. So much attention was focused on chitin starch [12,13]. Crosslinked starch xanthate and carboxyl-containing starch derivatives had been used as effective adsorbents to remove heavy metal ions from wastewater. Xu reported the mass transport process for the adsorption of chromium(VI) onto crosslinked amphoteric starch [14].

Konjac glucomannan (KGM) is a high molecular weight water-soluble non-ionic polysaccharide found in tubers of the amorphophallus konjac. KGM is a linear random copolymer of β-(1,4) linked D-glucose and D-mannose in the molar ratio of 1:1.6 with a low degree of acetyl groups [15,16]. It is more attractive for industrial use owing to its renewability, biodegradability and economy. At present, many investigations related to KGM are focused on isolating and characterizing properties of solution and solid, it is mainly used in biochemical, medical and food fields [17-20].

Recently, the investigation of carboxymethyl KGM used to adsorb Cu2+, Pb2+ and Cd2+ ions from the aqueous solutions was carried out by our group, the results showed KGM modified by anion etherifying agent can be used as an high effective and fast metal ion adsorbent. But the researches on KGM modified by cationic etherifying agent and its adsorption of heavy metal ions, especially Cr(VI) ion, have not been reported so far.

The objective of this study was focused on the development of konjac glucomannan modified by 3-chloro-2-hydroxypropyl-trimethylammonium chloride(HAT) and epichlorohydrin (ECH) for removal of Cr(VI)ions. The effects of pH, adsorbent dose, initial concentration of Cr(VI), adsorption time and temperature on adsorption capacity were investigated. In order to provide a new method and theoretical evidences for waste-water treatment, the characterization of isothermal adsorption and adsorption thermodynamics were also studied.

2. Materials and methods

2.1. Reagents and instruments

KGM powder was obtained from Shiyan Huaxianzi Konjac Productions Co., Ltd. Hubei Province, China and dried at 60°C in vacuum oven before used; HAT were provided by Dow Company Americas (aqueous solution with w(HAT)=69%), ECH and ethanol (analytical reagent grade) were obtained from Beijing Chemical Factory (Beijing, China). K2Cr2O7, NaOH, and other chemicals were of analytical reagents purchased from Beijing Chemical Factory (Beijing, China). US-VIS-NIR recording spectrophotometer (a model of UV-1700 made in Japan) was used to analyze the content of chromium(VI) in the aqueous solution.

2.2. Preparation of crosslinked cationic konjac glucomannan (CCKGM)

10g KGM was dispersed in 30 mL isopropyl alcohol and increased temperature to 40°C and kept constant, 15g NaOH solution (45wt.% in mass basis) was added slowly while stirring and certain amount of ECH was then added gradually at 40°C. After 1.0 h, the temperature was increased to 50°C, the defined HAT was dropped, and mechanical stirring was continued for 1.5 h at 50°C. The mixture was then allowed to cool, neutralized with hydrochloric acid, washed with 50 wt% alcohol to remove impurities, filtered and dried. The nitrogen content and the DS of quaternary ammonium cationic groups in the CCKGM were analyzed using element analysis (German elementar strucment). The degree of substitution (DS) and reaction yield (R) were calculated according to equations (1) and (2).

$$DS=\frac{162.15\times W(N)}{1401-151.64\times W(N)} \quad (1)$$

$$R=\frac{DS_{\text{test}}}{DS_{\text{theory}}}\times 100\% \quad (2)$$

During the experiment, only changing the content of cationic etherifying agent, three kinds of CCKGM with different DS were prepared and named CCKGM1, CCKMG2 and CCKGM3.

The preparation route was shown as Fig.1 and the preparation conditions and reaction results were shown in Tab. 1.

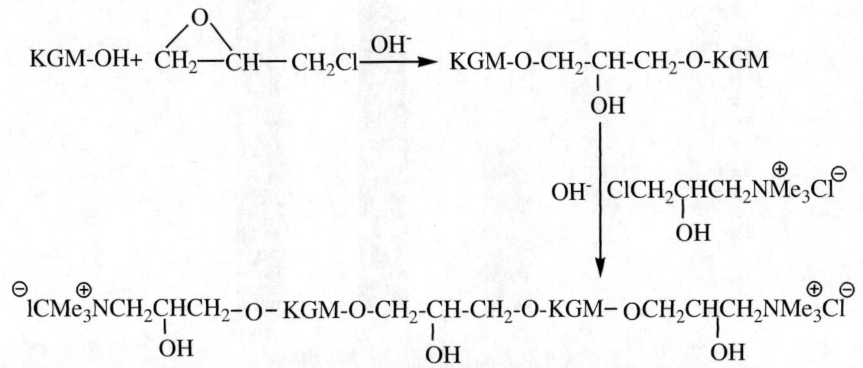

Figure 1. Preparation scheme of CCKGM

Table 1. Reaction conditions and reaction results of KGM with HAT.

Sample	T (°C)	Time (h)	n(KGM):n(HAT) :n(NaOH):n(ECH)	DS	R (%)
CCKGM1	50	1.5	1:0.8:1.5:1	0.226	28.3
CCKGM2	50	1.5	1:1.3:1.5:1	0.391	30.1
CCKGM3	50	1.5	1:1.0:1.5:1	0.456	45.6

2.3. Adsorption procedure

The adsorption experiments in this study were carried out according to the reference [14]. 50 mL of dichromate aqueous solution with desired concentration and the desired dose of CCKGM were placed in a series of comparison tubes in a thermostat water bath. Initial pH was adjusted with dilute HCl and 0.2 N NaOH aqueous solution. After shaken for the definite time, the tubes were removed and the concentration of Cr(VI) in the aqueous solution after the adsorption was analyzed by US-VIS-NIR recording spectrophotometer. The adsorption capacity of Cr(VI) on CCKGM was calculated by the equation (3).

$$Q = \frac{(C_0 - C_t)V}{m} \tag{3}$$

where C0 and Ct are the concentrations of Cr (VI) ions before and after adsorption respectively (mg/L); V is the total volume of the aqueous solution (L); and m is the dry weight of CCKGM (g).

3. Results and discussion

3.1. Comparison between native KGM and CCKGM in chromium(VI) adsorption

Native KGM itself appeared to adsorb minor amounts of chromium(VI) ions. As shown in Fig.2, adsorption capacity of native KGM was much lower in comparison with CCKGM. Adsorption capacity of native KGM was only 1.7 mg/g, while those of CCKGM1, CCKGM2, and CCKGM3 were 31.52 mg/g, 58.82 mg/g and 62.59 mg/g respectively. It may be explained that physical entrapment of the chromium(VI) ions would be the characteristics for the ion-adsorption of native KGM while strong electrostatic attraction would be the characteristics for the ion-adsorption of CCKGM.

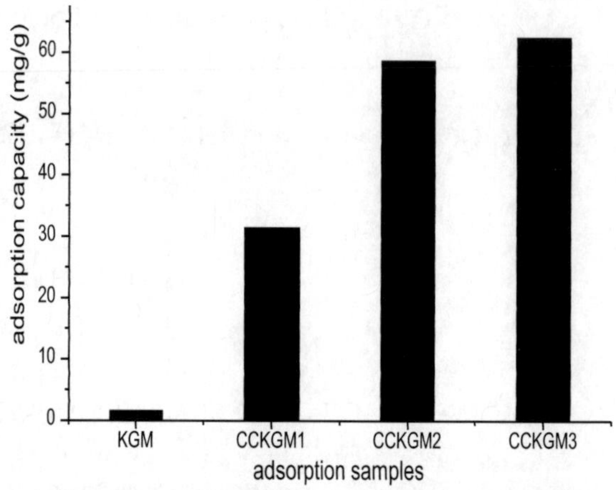

Figure 2. Effect of different adsorbent samples on the residual concentration of chromium(VI) solution (the dose of KGM and all CCKGM are 60mg; T =20 °C, pH=4, [Cr(VI)]=50mg/L, t=20min).

3.2. Isotherm adsorption

The Langmuir isotherm data together with the Freundlich data are often used to investigate adsorption behavior of heavy metal ions on polysaccharide. The Langmuir model is expressed as

$$\frac{C_e}{Q_e} = \frac{C_e}{Q_m} + \frac{1}{Q_m b} \tag{4}$$

where Ce is equilibrium concentration of the metal ion in aqueous solution (mg/L), Qe is adsorption capacity (mg/g), Qm is the maximum adsorption capacity (mg/g), and b is Langmuir constant (L/mg).

From the correlation coefficients listed in Tab.2, we could draw the conclusion that the adsorption of Cr (VI) ions onto CCKGM was well followed as Langmuir isotherm equation under the concentration range studied. It indicated the monolayer coverage of CCKGM by the ions, and all the adsorption sites supplied nearly same action to metal ions. This could reflect indirectly that ligand ion exchange was predominant in the adsorption process. A higher DS value of the cationic group led to a higher Qm and b, it corresponded to the above-mentioned conclusion that a higher DS resulted in a higher adsorption capacity.

Table 2. Langmuir parameters for CCKGM1, CCKGM2 and CCKGM3

	Qm(mg/g)	b(L/g)	R*
CCKGM1	41.84	0.084	0.9892
CCKGM2	67.79	0.442	0.9941
CCKGM3	77.38	1.615	0.9922

Note: R * is the correlation coefficient of Langmuir isothermal adsorption plots.

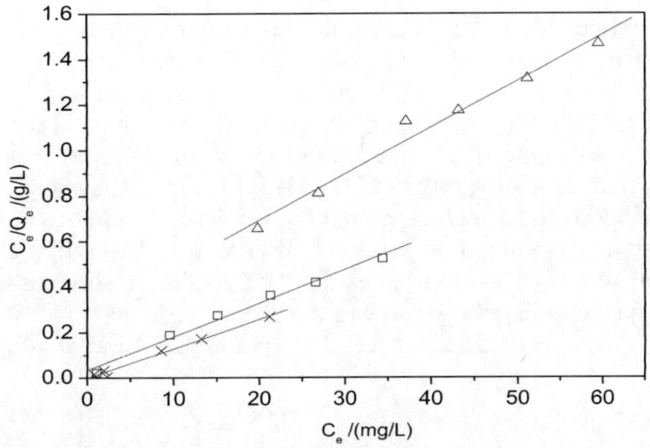

Figure 3. Langmuir adsorption isotherm for CCKGM1, CCKGM2 and CCKGM3: (△)CCKGM1; (□)CCKGM2; (X) CCKGM3 (pH=3.78; t=1 h; T=20 °C; the dose of CCKGM1, CCKGM2, CCKGM3 each was 50 mg).

The essential features of a Langmuir isotherm can be expressed in terms of a dimensionless constant separation factor or equilibrium parameter, RL, which is used to predict if an adsorption system is "favorable" or "unfavorable". The separation factor, RL is defined by:

$$R_L = \frac{1}{1+bC_0} \quad (5)$$

where C0 is the initial concentration of metal ions (mg/L), and b is the Langmuir adsorption equilibrium constant (mL/mg). Because b values were all beyond zero, RL values were in the range 0＜RL＜1, which indicated that the adsorption of Cr (VI) ions of CCKGM was favorable. Thus, CCKGM was favorable adsorbent for Cr (VI) ions.

3.3. Effect of adsorption time on residual concentration of Cr(VI)

Fig.4 showed that the time profiles of residual concentration of chromium(VI) changed with the initial concentration of 60 mg/L. It was clear that time had evident influence on the residual concentration of chromium(VI) solution within 20 min. However, the decrease in the residual concentration of chromium(VI) solution was very small beyond 20 min and was independent of DS. In process application, the rapid adsorption phenomenon was an advantage because short adsorption time could directly affect operation cost.

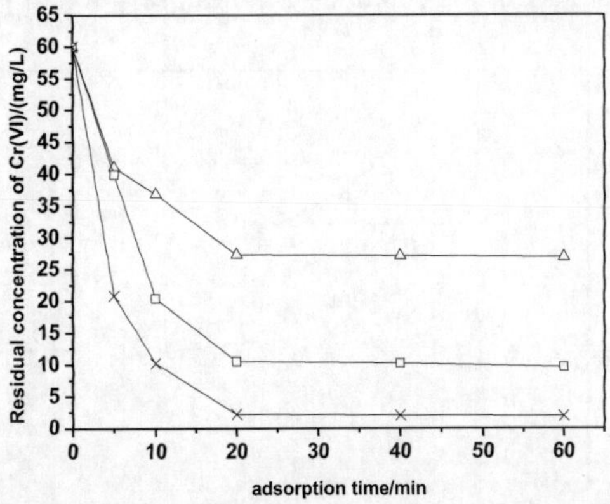

Figure 4. Effect of adsorption time on residual concentration of Cr(VI) (△)CCKGM1; (□)CCKGM2; (×)CCKGM3 (pH=3.78; T=20 °C; the dose of CCKGM1, CCKGM2, CCKGM3 each was 50 mg; the initial concentration of Cr(VI) was 60 mg/L).

IOP Conf. Series: Materials Science and Engineering **587** (2019) 012010 doi:10.1088/1757-899X/587/1/012010

3.4. Effect of adsorbent dose

Fig.5 showed the effect of adsorbent dose on adsorption capacity under the condition of pH=3.78, C0=100 mg/L and t=60 min. The adsorption capacity of Cr(VI) adsorbed by different CCKGM followed the same trend, that is, adsorption capacity decreased with increasing of CCKGM dose in the range of 30 mg to 80 mg, which is contrary to the theory analysis. The reason was the quantity of Cr(VI) ions around activity sites of the single adsorbent decreased as the concentration of Cr(VI) kept constant, namely, the contact chance of CCKGM active site with Cr(VI) ion decreased. Because adsorption capacity is defined as the metal ion mass (mg) adsorbed by per gram adsorbent, so adsorption capacity decreased. As far as improving Q was concerned, increasing the DS of the absorbents was more effective and economical than increasing the absorbents dose.

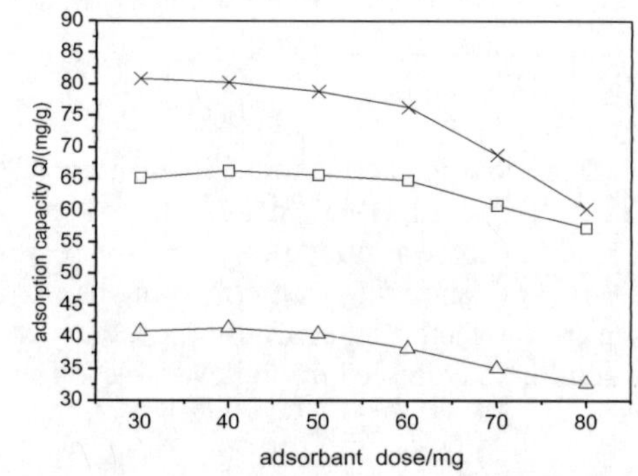

Figure 5. Effect of dose on adsorption capacity for CCKGM1,CCKGM2 and CCKGM3:(△) CCKGM1; (□)CCKGM2; (×) CCKGM3 (pH=3.78, t=1 h, T =20 °C); the initial concentration of Cr(VI) was 100 mg/L).

3.5. Effect of pH on the adsorption

Metal-ion adsorption was known to be dependent on the pH of solution (see Fig.6). The adsorption capacity changed little with pH from 3.0 to 4.0, while decreased significantly as the pH increased from 4 to 10, because of the reaction of equation (6) in the aqueous solution.

$$Cr_2O_7^{2-}+2OH^- \rightleftharpoons 2CrO_4^{2-}+H_2O \qquad (6)$$

In acid media, Cr2O72- is the main form, the amount of CrO_4^{2-}groups in aqueous solution tended to increase when pH values of Cr(VI) solution increased and CrO_4^{2-} is the main form. The sum of active sites was fixed, and electrostatic attraction was the main effect between CCKGM and Cr(VI). The adsorption capacity for Cr2O72- was stronger than that for CrO_4^{2-}on CCSM. So the optimal pH condition for the adsorption of Cr(VI) was about 4.0.

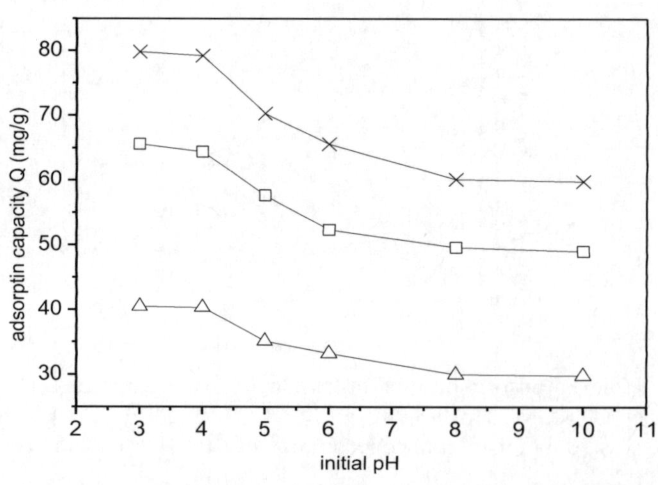

Figure 6. Effect of initial pH of solution on adsorption capacity for CCKGM1, CCKGM2 and CCKGM3:(△) CCKGM1;(□) CCKGM2; (×) CCKGM3 (t=1 h; T=20°C; the dose of CCKGM1, CCKGM2, CCKGM3 each was 50 mg; the initial concentration of Cr(VI) was 100 mg/L)

3.6. Effect of the Cr(VI) initial concentration

In general, the concentration of Cr(VI) in industry wastewater is lower than 100 mg/L. So the effect of Cr(VI) initial concentration in the range 50~100 mg/L on adsorption capacity were studied under the condition of T=20°C, pH=3.78, m=50 mg/L and t=60 min. The results showed in Fig.7 indicated that the adsorption capacity was directly influenced by Cr(VI) initial concentration and increased with Cr(VI) initial concentration. As the initial concentration was 100 mg/L, the adsorption capacity of three kinds of CCKGM was respectively 40.5 mg/g, 65.6 mg/g and 78.8 mg/g, higher than 30.2 mg/g, 49.2 mg/g and 48.9 mg/g with the initial concentration 50 mg/L. The active sites in CCKGM were more than Cr(VI) with the low Cr(VI) concentration. While as the concentration was high, there were enough Cr(VI) contacted with CCKGM active sites, which resulted in adsorption capacity high, but the proportion of Cr(VI) unabsorbed increased.

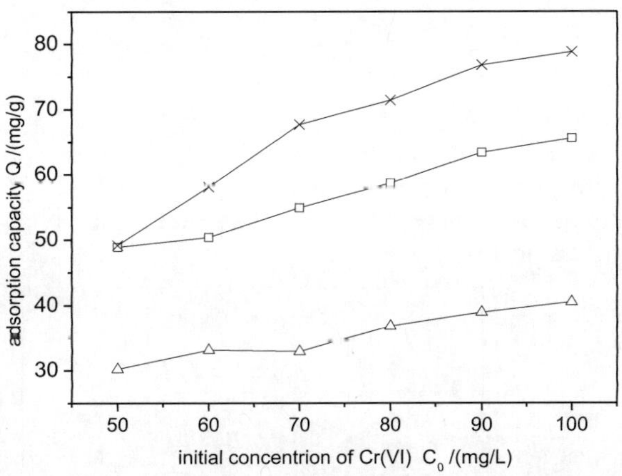

Figure 7. Effect of initial concentration of Cr(VI) on adsorption capacity for CCKGM1, CCKGM2 and CCKGM3(△) CCKGM1; (□)CCKGM2; (×) CCKGM3 (pH=3.78; t=1 h; T=20 °C; the dose of CCKGM1, CCKGM2, CCKGM3 each was 50 mg).

3.7. Thermodynamic studies

Effect of temperature on the Cr(VI) adsorption capacity is shown in Fig.8. The Cr(VI) adsorption capacities with various temperatures (20–60°C) were determined when initial solution concentration and pH were 60 mg/L and 3.78, respectively, and the dose of each CCKGM was 50 mg. The results showed that the adsorption capacity decreased with increasing temperature, which implied that the adsorption process was exothermic. For verification of the conclusion, the curves of log (Q/Ce) versus 1/T for CCSM1, CCSM2, and CCSM3 were shown in Fig.9. the results of ΔH and ΔS calculated according to the equation(7) and ΔG according to the equation (8) were listed in Tab.3.

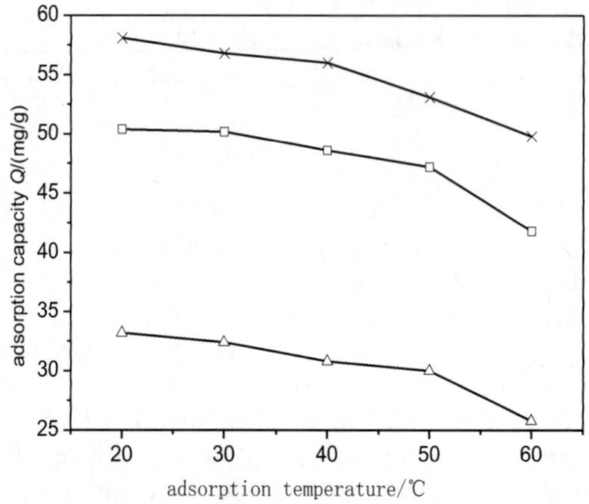

Figure 8. Effect of temperature on adsorption capacity. (△)CCKGM1; (□)CCKGM2; (×)CCKGM3

$$\lg(\frac{Q}{C_e}) = -\frac{\Delta H}{RT} + \frac{\Delta S}{R} \quad (7)$$

$$\Delta G = \Delta H - T\Delta S \quad (8)$$

The negative value of ΔH indicated that the adsorption process was exothermic and low temperature made the adsorption easy. Moreover, the negative values of ΔS and ΔG demonstrated a decrease in adsorption energy and an increase in the feasibility of adsorption process at lower temperature.

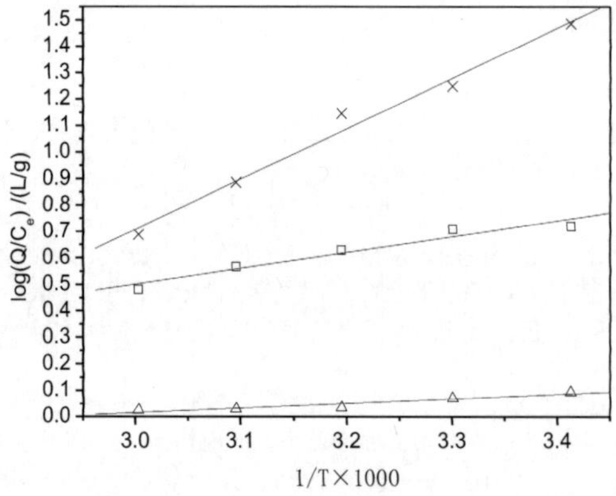

Figure 9. Plot of 1/T×103 versus log(Q/Ce) for (△)CCKGM1; (□)CCKGM2; (×)CCKGM3 (pH=3.78; t=1 h; the initial concentration of Cr(VI) was 60 mg/L; the dose of CCKGM1, CCKGM2, CCKGM3 each was 50 mg).

Table 3. Results of entropy (△S), free energy (△G) and enthalpy (△H) of three kinds of CCKGM

sample	T (°C)	Q (mg/g)	△G (kJ/mol)	△H (kJ/mol)	△S (J/mol)
CCKGM1	20	33.2	-0.212	-1.46	-4.26
	30	32.4	-0.169		
	40	30.8	-0.127		
	50	30.0	-0.084		
	60	25.4	-0.041		
CCKGM2	20	50.4	-1.827	-5.00	-10.83
	30	50.2	-1.719		
	40	48.6	-1.610		
	50	47.2	-1.502		
	60	41.8	-1.394		
CCKGM3	20	58.1	-3.642	-15.84	-41.63
	30	56.8	-3.226		
	40	56.0	-2.810		
	50	53.1	-2.394		
	60	49.8	-1.977		

4. Conclusion

Cationic konjac glucomannan could effectively remove Cr(VI) ions from water solutions. Adsorption capacity was affected by solution pH, initial concentration of metal ions, adsorbent dose and DS of cationic group. The adsorption capacity increased with DS and the value was respectively 31.52 mg/g, 58.82 mg/g and 62.59 mg/g corresponding to the DS 0.226,0.391 and 0.456. The adsorption process could be finished within several minutes and followed Langmuir isothermal adsorption under the concentration range studied. The process was exothermal and low temperature was favorable for adsorption. CCKGM had potential to be used as absorbent for Cr (VI).

References

[1] Ma Islam, Liu h., Zhao,J., Sun et al.Selective and efficient removal of toxic omoanions of As(Ⅲ) and Cr(Ⅵ) by layered double hydroxide intercalated with MoS42-, Chemistry of Materials. 29(2017)3274-3284

[2] G. McMullan, C. Meehan, A. Conneely, N. Kirby, T. Robinson, P. Nigam, Microbial decolourisation and degradation of textile dyes, Appl Microbiol Biotechnol. 56 (2001) 81-87.

[3] Li,M.,Lv,Z., Zheng,J. Positively charged nanofiltration membrane eith dendritic surface for toxic element removal. ACS Sustainable Chemistry & Engineering, 5(2017)784-792

[4] Tan,J.Z.,Xia,F.,SaniM.A. High performance coral reef-like carbon nitrides: Synthesis and application in photocatalysis and heavy metal ion adsorption. ACS Applied Materials & Interfaces. 9(2017)4540-4547

[5] G.U. Von, Part 1 Oxidation kinetics and product formation, Water Res. 37 (2003) 1443-1467.

[6] Heffron,.J.,Marhefke,M.,& Mayer,B.K. Removal of trace metal contamination from potable water by electrocoagulation. Scientific Reports,2016,6,28478.

[7] Z. Hu, L. Lei, Y. Li, Y. Ni, Chromium adsorption on high-performance activated carbons from aqueous solution, Sep Purif Technol. 31 (2003) 13-18.

[8] Z. Reddad, C. Gerente, Y. Andres, J.F. Thibault, P. Le Cloirec, Cadmium and lead

adsorption by natural polysaccharide in MF membrane Reactor: experimental analysis and modeling, Water Res. 37 (2003) 3983-3991.

[9] A.J. Varma, S.V. Deshpande, J.F. Kennedy, Metal complexation by chitosan and its derivatives: a review, Carbonhydr Polym. 55 (2014) 77-93.

[10] W. Ciesielski, C.Y. Lii, M.T. Yen, P. Tomasik, Interaction of starch with salts of metals from the transition groups, Carbohydr Polym. 51 (2003) 47-56.

[11] E. Polaczek, F. Starzyk, K. Malenki, P. Tomasik, Inclusion complexes of starches with hydrocarbons, Carbohydr Polym. 43 (2000) 291-297.

[12] M.N.V. Ravi Kumar, A review of chitin and chitosan applications, React Funct Polym. 46 (2000) 1-27.

[13] S. Babel, T. A. Kurniawan, Low-cost adsorbents for heavy metals uptake from contaminated water a review, J Hazardous Mat. B97 (2003) 219-243.

[14] S.M. Xu, S.F. Zhang and R.W. Lu, Study on adsorption behavior between Cr(VI) and crosslinked amphoteric starch, Journal of Applied Polymer Science 89 (2003), pp. 263–267.

[15] M. Maeda, H. Shimahara, N. Sugiyama, Detailed examination of the branched structure of konjac glucomannan, Agric, Biol. Chem. 38 (1980) 315-323.

[16] F. Smith, H.C. Srivastava, Constitution studies on the glucomannan of konjac flower, Journal of American Chemical Society 39 (1959) 1715-1718.

[17] C. Huang, Effect of konjac food on the blood glucose level in diabetics. Acta Nutuitional Sinico 2 (1990) 360-366.

[18] K. Hannigan, This food helps control weight. Food Engineering International 5(1980) 21-25.

[19] Y. Gu, J. Shi, Z. Hu, Fight against senium action of konjac polysaccharides. Chin Tradit Herb Drugs 30 (1999) 78-81.

[20] C. Mao, N. Yu, Z. Gu, Hypoglycemic effects of amorphophallus konjac in alloxan diabetic rats. Chin Mod Appl Pharm 18 (2001) 12-17.

IOP Conf. Series: Materials Science and Engineering **587** (2019) 012011 doi:10.1088/1757-899X/587/1/012011

Mechanical properties and crack resistance of sustainable high strength engineered cementitious composites

Y D Han[1], Q Wang[2*], Z B Wang[3] and B Y Yao[2]

[1]Central Research Institute of Building and Construction, MCC Group Cooperation Limited, Beijing 100088, China
[2]Department of Civil Engineering, Guangzhou University, Guangzhou 510006, China
[3]School of Mechanics and Civil Engineering, China University of Mining and Technology, Beijing 100083, China

Corresponding author and e-mail: Q Wang, wangq@gzhu.edu.cn

Abstract. Engineered cementitious composite (ECC) is a class of high performance fiber reinforced material designed to overcome the brittleness of concrete. The high content of cementitious material restricts the common application of ECC. To develop a new series of sustainable and durable composite, industrial wastes such as fly ash, silica fume and furnace slag were brought in as supplementary cementitious materials (SCMs) in present study. High strength cement matrixes with three types of SCMs were obtained. Tensile performance of PVA-steel fiber reinforced engineered cementitious composite based on above matrixes were evaluated via uniaxial tensile test. Crack resistance of high strength engineered cementitious composite (HSECC) with industrial wastes were experimentally estimated through steel ring test. Results show that suitable addition of industrial wastes somchow improved the mechanical performance and crack resistance of HSECC and further enhanced the durability and sustainability.

1. Introduction

Concrete is a typical brittle material, as first crack occurs, immediate localization of deformation appears followed by sudden decreasing load. Cracks in concrete component often threaten the durability and cause catastrophic damage of building structures. To overcome the brittle nature of concrete, a class of high performance ductile cementitious composite, called Engineered Cementitious Composites (ECC), has been developed recently [1-2]. It has a unique property of strain-hardening accompanied by multiple cracking. Ductility of ECC is resulted from the additional fine cracks formed under tension. Typically, strain localization occurs at a tensile strain of 1-3%, with crack spacing of 3-6mm and crack width around 60μm [3-5]. Such capillary cracks will have little effect on the permeability of water and other deleterious substance. As a consequence, with slight degradation in carrying capacity and transport properties under high deformation, the durability of concrete structure can be maintained [6].

However, in order to obtain this ductile and multiple cracking behaviors, coarse aggregates shall be eliminated and the content of cementitious material increases correspondingly. Generally, admixture of ECC contains over 1000 kilogram in 1 cubic meter. Considering equal-mass carbon dioxide released in the process of cement producing, traditional ECC with cement only as binding material is definitely not sustainable. Researches were carried out in the past decade to develop eco-friendly ductile materials, and industrial wastes such as fly ash from power station[7-9], silica fume collected in metallurgical industry [10], rice husk from burnt straws[11], ferrous tailings released by iron industry [12-13] and ground granulated blast furnace slag from steel works[14-16], were introduced to the cementitious system of ECC. Since the presence of above powders, flowing ability and mechanical properties of ECC altered significantly due to their supplementary cementitious function. For example, vitreous particles of fly ash enable cement granules to distribute more homogeneously and workability of cementitious materials is improved simultaneously. Industrial by-products provide possibilities for development of green ECC.

On the other hand, constructions indicate that compressive strength of cement matrixes used in green ECC are usually lower than 50 MPa [3, 6]. Even though the ability of crack width controlling maintains adequately, a number of applications, such as permanent formwork for concrete structures, steel-concrete composite bridge decks [17-19], still need the ductile material possesses a relatively higher strength. Apart from strain-hardening characteristic, material properties with high crack resistance ability are required as well. Therefore, high strength ductile cement matrixes incorporating industrial wastes for a new generation of ECC focusing on above applications are in essential need of developing.

In the present paper, high strength engineered cementitious composite (HSECC) is designed using three kinds of industrial waste as supplementary cementitious materials (SCMs). The compressive strength and bending strength of high strength cement matrixes are experimentally evaluated. Tensile performance of HSECC with polyvinyl alcohol and steel fiber is investigated through uniaxial tension tests. In addition, steel ring tests were carried out to obtain the crack resistance ability of HSECC.

2. Experimental program

The purpose of this test program is to investigate the effect of industrial by-products on compressive and bending strengths of high strength cement matrix, tensile performance of HSECC and to obtain the crack resistance behaviour. To achieve this target, three kinds of industrial waste, silica fume (SF), grounded granulated blast furnace slag (SG) and fly ash (FA) were used in the experiments. Silica sands with particle size of 0.075 to 0.150mm, polyvinyl alcohol (PVA) and steel fiber (ST) were used in the tests.

2.1. Materials

The chemical compositions of ordinary Portland cement (OPC) and the three SCMs (SF, SG, FA) used in tests are listed in Table 1. The particle size distribution of above cementitious materials is shown in Figure 1. Fibers used to evaluate the tensile performance of ECC with the developed high strength matrix include Polyvinyl alcohol fiber (PVA) supplied by Kuraray company from Japan and steel fiber supplied by Changhong company in China. The mechanical properties of the fibers are listed in Table 2. In the mix design of ECC, 2% volume fraction of PVA fibers and 1% volume fractions of steel fibers were applied.

Table 1. Chemical composition of the cementitious materials used in tests (wt%)

No	SiO_2	CaO	Al_2O_3	Fe_2O_3	MgO	K_2O	Na_2O	SO_3	LOI
OPC	23.67	59.98	7.21	3.07	2.07	0.62	0.17	2.14	1.01
Silica fume (SF)	90.56	0.81	0.41	0.52	0.95	1.59	0.63	-	3.72
Slag (SG)	38.83	38.70	12.92	1.46	4.63	0.37	0.28	0.60	0.06
Fly ash (FA)	47.02	5.08	35.06	3.88	1.36	1.30	1.18	0.89	1.85

2.2. Mixture, specimens and curing

Tested mix proportion incorporating with SF, SG and FA are displayed in Table 3. Almost 30 percent of cement is replaced by industrial waste. Rectangular specimens with size of 40×40×160 mm was

IOP Conf. Series: Materials Science and Engineering **587** (2019) 012011 doi:10.1088/1757-899X/587/1/012011

cast to conduct the compressive and bending tests for each mixture at age of 3, 7 and 28 days. For uniaxial tensile test, rectangular coupon specimens with size of 100×200×20 mm were used. After removing from their molds, specimens were stored in water at 20±2°C for curing. The tensile experiments were tested under uniaxial tension with displacement control of 0.0025 mm per second in a 250 kN capacity MTS 810 material testing system. Aluminum plates were epoxy glued onto the ends of the specimens prior to loading at least 6 hours to enhance the ends for gripping. Tensile strain was measured by two extensometers mounted on the lateral surface of the coupon specimen. The measured gage length of extensometer was 50 mm, as shown in Figure 2.

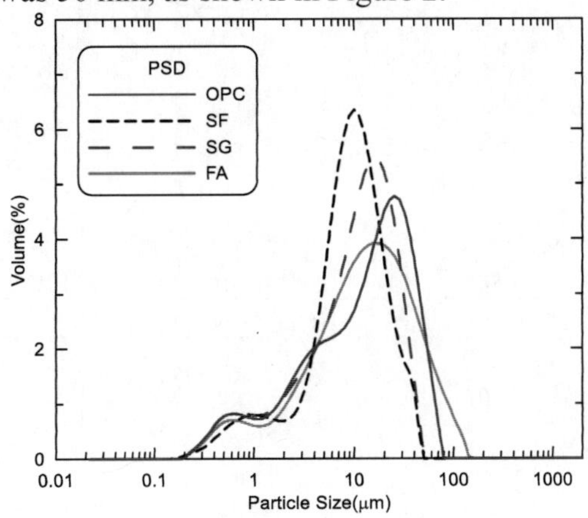

Figure 1. Particle size distribution of the five fine cementitious materials

Table 2. Properties of PVA and steel fibers

Fiber type	Density (g/cm^3)	Tensile strength (MPa)	E (GPa)	Diameter (mm)	Length (mm)
PVA	1.2	1620	42.8	0.039	12
ST	7.8	2750	210	0.200	13

Table 3. Mix proportion of tested high strength cement matrixes

Test Series OPC	Cementitious Materials Total(Cement:SF:SG:FA)	Quartz Sand	Water	Superplasticizer (%)
SF	1(0.90:0.10:0:00:0.00)	0.833	0.192	1.79
SG	1(0.80:0.10:0.10:0.00)	0.833	0.195	1.95
FA	1(0.70:0.10:0.10:0.10)	0.833	0.181	1.32

2.3. Testing procedures

The mixing procedure of the composite material consists of mixing cementitious materials, adding water with superplasticizer and spreading fibers. To evaluate the crack resistance ability of HSECC with industrial waste, ring test was adopted. To obtain the actual composite-steel interfacial compressive stress induced by shrinkage of ECC, four strain gauges along the inner wall of steel ring are placed according to Figure 3(a). A steel ring with the outer diameter of 320 mm and the inner diameters of 290 mm was used in the tests. The thickness of composite was 35 mm and the height 150 mm. The geometry of the ring specimen used in the present study is presented in Fig.3. Each ring specimen was equipped with four strain gauges with a length of 3 mm and width of 2 mm, placed at mid-height on the inner circumference of the steel ring. The fresh composite was cast into the ring mold in two layers and was consolidated by a vibrating table. As is shown in Figure 3(b), drying direction makes specimen exposed to the air and accelerate the process of cracking. Ring test is proved to be a simple but effective method to estimate the anti-cracking ability of cement-based materials[20].

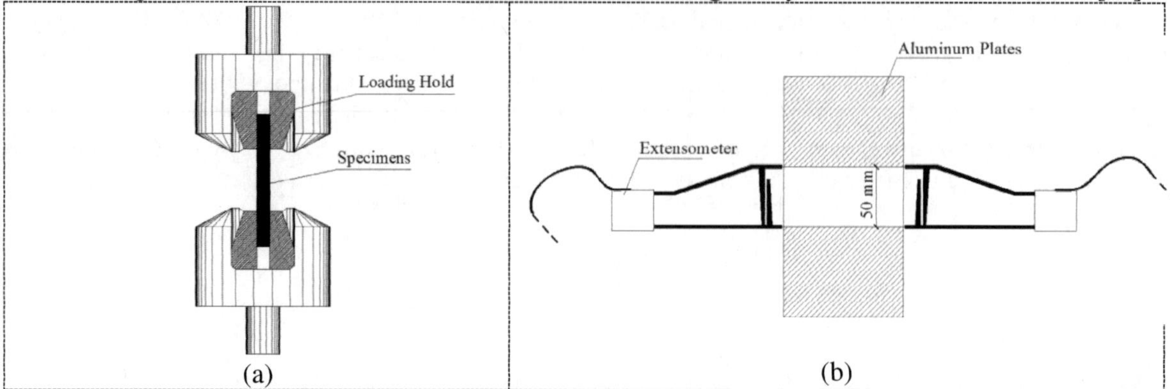

Figure 2. (a)Tensile test set-up (b) specimen with aluminum plates glued and extensometers mounted

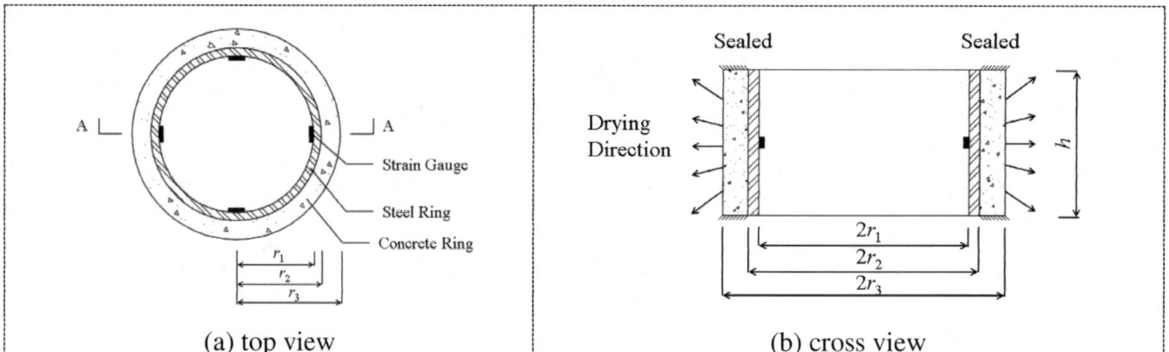

Figure 3. Geometry of the ring specimen

3. Results and Discussion

Mechanical properties and crack resistance behavior of sustainable high strength ductile cementitious composites with SF, SG and FA as supplementary cementitious materials were experimentally studied. Results were as follows.

3.1. Bending and compressive strength

The overall enhancing efficiency on compressive and bending strength of high strength cement matrix by using silica fume, slag and fly ash is graphically displayed in Figure 4, in which average strength at 3, 7 and 28 days in each test series, including the reference, is presented. Obviously, through the optimizing procedure, compressive and bending strength of the low water to binder ratio matrix can significantly be increased by substituting cement with SF, SG and FA respectively. For example, with a similar fluidity of the fresh mortar, the compressive strength at 3, 7 and 28 days changes from 45.0, 57.8 and 77.9 MPa to 62.0, 80.6 and 108.0 MPa. And the corresponding bending strength at 3, 7 and 28 days may change from 8.1, 12.4 and 14.7 MPa to 10.9, 15.2 and 18.7 MPa. The utilization of industrial wastes not only lower the amount of cement and cut down the contamination, but also enhance the strength of ductile cementitious materials. General range of strength increasing is about 20% to 70% in comparison with the reference. It is rather interesting that such significant strength increasing is based only on the optimization on composition of cementing binder with SF, SG and FA, those actually are industrial by-products and the cost is lower than cement in general, but their filler action and particle packing effect make a great contribution to reduce the porosity of capillary pores in the high strength matrix and minimize the initial crack inside the composite.

IOP Conf. Series: Materials Science and Engineering **587** (2019) 012011 doi:10.1088/1757-899X/587/1/012011

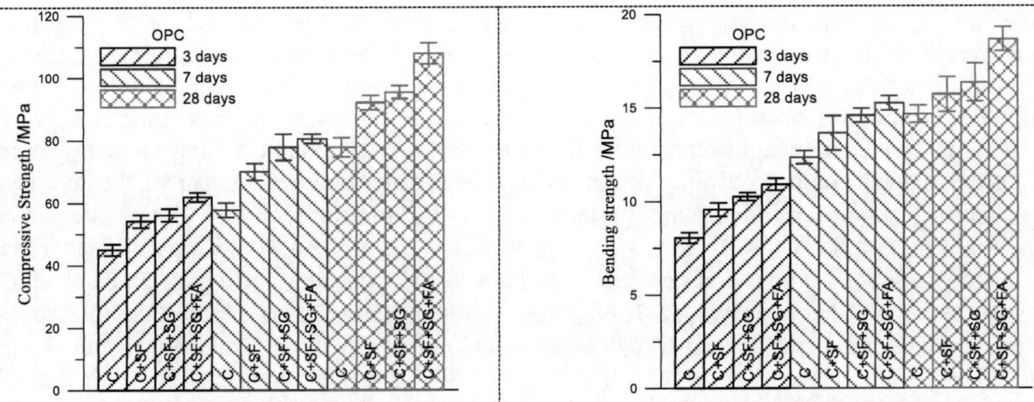

(a) Compressive strength (b) Bending strength

Figure 4. Compressive and bending strength of ECC with SF, SG and FA

3.2. Tensile performance

Tensile performance of PVA-steel fiber reinforced cementitious composites based on the developed high strength matrix is preliminary evaluated by uniaxial tension tests. Figure 5 displays the tensile stress-strain curves of the composites at the age of 7 and 28 days respectively. From the results, we may find that the strain-hardening performance is largely enhanced compared to cement mortar without ductile fibers. The average tensile strength and strain are 7.27MPa, 0.688% and 7.49MPa, 0.815% respectively at 7 and 28 days. Generally, material with tensile strain capacity of 0.5% is ductile for concrete structure. From the tensile curve, fluctuations occur at the level of stress capacity which may indicate the crack propagation process and each crest represents one capillary crack. As a consequence, HSECC with industrial waste maintains the properties of strain-hardening and multiple cracking. In addition, strength is a critical parameter that controls the penetration of chemical agent during service of the structures at the stage without cracking. Strain capacity behaving as the formation of fine cracks is more important for the long term durability and sustainability of the structures at the stage that cracks may occur. The balance optimization between strength, strain and crack resistance of ECCs are definitely needed to meet the requirements of practical applications. Therefore, the crack resistance of PVA-steel hybrid fiber reinforced ECC with developed high strength matrix is experimentally investigated in the following.

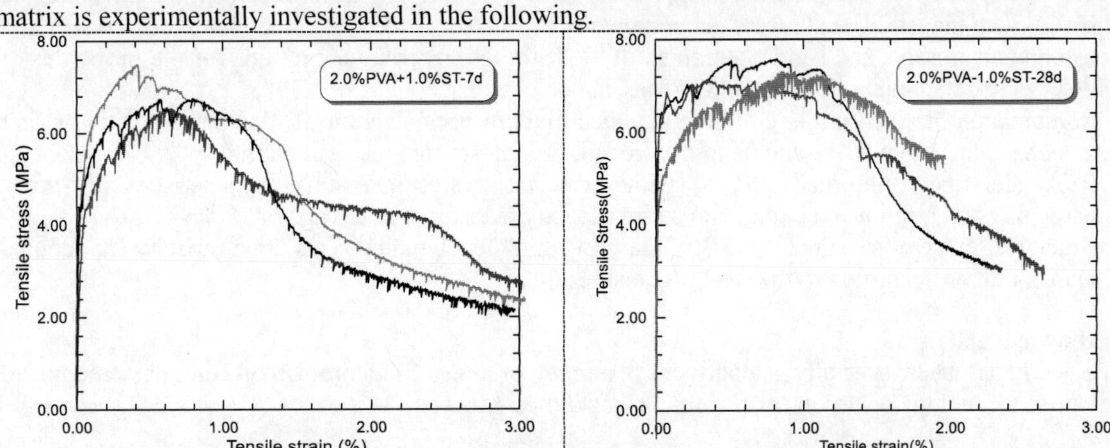

(a) 7days (b) 28days

Figure 5. Tensile stress-strain curves at 7 and 28 days

3.3. *Cracking resistance*

To obtain anti-cracking ability of the materials under tensile load, steel-HSECC composite ring tests were conducted. From the ring test, we can obtain the progress of compressive stress in the steel ring, which result from shrinkage of the composite ring. Figure 6 displays the typical curves of hoop strain as a function of curing time obtained by strain gauge attached to the inner circumferential steel ring. Increase of compressive strain can be seen after the initial steady stage, and cracking moment can be deduced by the sudden climb of strain. As for cementitious materials, hoop strain of the interface between steel and composite rings stays around -90 to -120 μm/m. Figure 7 shows the actual circumferential strain of HSECC with SCMs and hybrid fibers. From the curves we may observe that hoop strain increases along with the curing period and reaches -39 μm/m at the age of 28 days which indicates that there is another approximate 51-81 μm/m strain capacity for external load applied to the composite in order to cause cracking. In addition, increasing rate of the strain demonstrates a declining trend and tends to be stable after certain curing time which provides a further proof that crack resistance of ductile cementitious material is somehow enhanced.

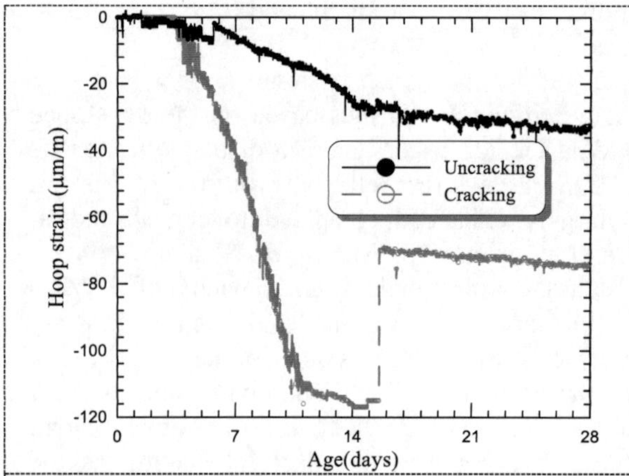

Figure 6. Typical ring test results

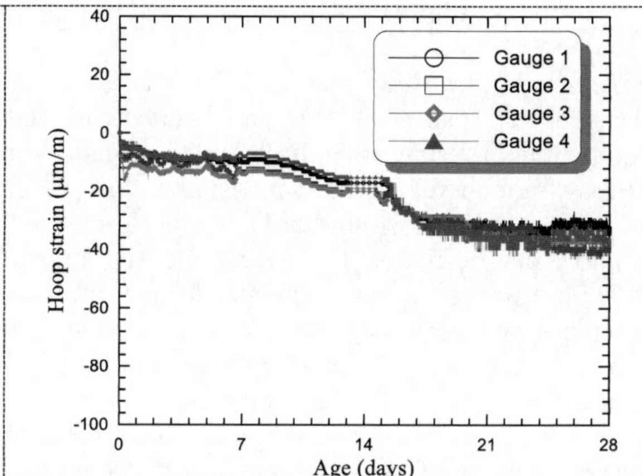

Figure 7. Hoop strain of HSECC

4. Conclusions

Mechanical properties and crack resistance of high strength engineered cementitious composite (HSECC) with industrial waste were experimentally evaluated. Effects of silica fume, furnace slag and fly ash on compressive and bending strength of high strength matrix were tested. Tensile properties of HSECC were obtained. Following conclusions can be drawn:

- Combination of silica fume, ground granulated blast furnace slag and fly ash has a positive effect on enhancing compressive and bending strength of high strength cement matrix.
- PVA-steel fiber reinforced HSECC with industrial wastes maintains characteristics of strain-hardening and multiple cracking with strain capacity over 0.8% at the age of 28 days .
- Crack resistance of sustainable HSECC stays in a high level and presents large capacity for coupled load condition according to the steel ring test results.

Acknowledgement

This work has been specially supported by a grant of open fund project of National recognized enterprise technology center in MCC Group Cooperation Limited.

References

[1] Li VC 2002 Advances in ECC Research, ACI Special Publication on Concrete: *Material Science to Applications* **SP 206-23** pp 373-400.

[2] Li VC 2007 Progress and application of engineered cementitious composites *Journal of the*

Chinese Ceramic Society **35(4)** pp 531-536.

[3] Li VC 1993 From Micromechanics to structural engineering--the design of cementitious composites for civil engineering applications *JSCE Journal of Structural Mechanics and Earthquake Engineering* **10(2)** pp 37-48.

[4] Yang EH and Li VC 2014 Strain-rate effects on the tensile behavior of strain-hardening cementitious composites. *Construction and Building materials* **52** pp 96-104.

[5] Lu C and Leung CKY 2017 Theoretical evaluation of fiber orientation and its effects on mechanical properties in Engineered Cementitious Composites (ECC) with various thickness. *Cement and Concrete Research* **95** pp 240-246.

[6] Zhang J, Wang Q and Wang ZB 2016 Optimizing design of high strength cement matrix with supplementary cementitious materials *Construction and Building Materials* **120** pp 123-136.

[7] Mustafa S and Li VC 2009 Durability properties of micro-cracked ECC containing high volumes fly ash *Cement and Concrete Research* **39** pp 1033-1043.

[8] Qian SZ and Zhang ZG 2012 Development of engineered cementitious composites with local ingredients *Journal of Southeast University* **28** pp 327-330.

[9] Ozbay E, Karahan O, Lachemi M, Hossain KMA and Duran C 2012 Investigation of properties of engineered cementitious composites incorporating high volumes of fly ash and metakaolin *ACI Materials Journal* 2012 pp 565-571.

[10] Zhang J, Wang Q and Wang ZB 2017 Properties of polyvinyl alcohol-Steel hybrid fiber-reinforced composite with high-strength cement matrix *Journal of Materials in Civil Engineering* **29** pp 1-13.

[11] Debora PR, Fernanda BPC, Angela GG and Luiz CPSF 2017 Tensile behavior and durability issues of engineered cementitious composites with rice husk ash *Revista Materia* **22.**

[12] Huang XY, Ranade R, Ni W and Li VC 2013 Development of green engineered cementitious composites using iro ore tailings as aggregate *Construction and Building Materials* **44** pp 757-764.

[13] Huang XY, Ranade R and Li VC 2013 Feasibility study of developing green ECC using iron ore tailings powder as cement replacement *Journal of Materials in Civil Engineering* **25** pp 923-931.

[14] Qiu JS, Tan HS and Yang EH 2016 Coupled effects of crack width, slag content, and conditioning alkalinity on autogenous healing of engineered cementitious composites *Cement and Concrete Composites* **73** pp 203-212.

[15] Khan MI, Fares G and Mourad S Optimized fresh and hardened properties of strain hardening cementitious composites: Effect of mineral admixture, cementitious composition, size, and type of aggregate *Journal of Materials in Civil Engineering* **29** pp 1-16.

[16] Chen ZT, Yang YZ and Yao Y 2013 Quasi-static and dynamic compressive mechanical properties of engineered cementitious composite incorporating ground granulated blast furnace slag *Materials and Design* **44** pp 500-508.

[17] Shao X, Yi D, Huang Z, Zhao H, Chen B, and Liu M 2013 Basic performance of the composite deck system composed of orthotropic steel deck and ultra thin RPC Layer *Journal of Bridge Engineering* **18(5)** pp 417-428.

[18] Zhang J, Wang Z, Ju X and Shi Z 2014 Simulation of flexural aperformance of layered ECC-concrete composite beam with fracture mechanics model *Engineering Fracture Mechanics* **131** pp 419-438.

[19] Wang Z, Zhang J, Wang J and Shi Z 2015 Tensile performance of polyvinyl alcohol-steel hybrid fiber reinforced cementitious composite with impact of water to binder ratio. *Journal of Composites* **49(18)** pp 2169-2186.

[20] Wang Q, Zhang J, Fan JS and Zheng X 2018 Cracking load of high strength ECC-steel composite bridge deck with impact of shrinkage stress-Test and simulation *Engineering Fracture Mechanics* **202** pp 174-201.

IOP Conf. Series: Materials Science and Engineering **587** (2019) 012012 doi:10.1088/1757-899X/587/1/012012

Metal fault analysis and preventive measures for grounding accessories of 110kV cable outdoor terminal

Y Xie[1*], L Yang[2], W Li[3] and Y L Liu[1]

[1] State Grid Hunan Electric Power Corporation Research Institute, Changsha, Hunan, 410007, China
[2] State Grid Xiangtan Power Supply Company, Xiangtan, Hunan, 411104, China
[3] State Grid Hengyang Power Supply Company, Hengyang, Hunan, 421000, China

Corresponding author and e-mail: Y Xie, yxie001@126.com

Abstract. There were three grounding metal faults of the 110 kV outdoor cable terminals, i.e., cracked weld between grounding copper block and the cable tail pipe, no fastened grounding copper block connecting bolt and no fixed grounding wire connected to the grounding copper block, respectively. Precautionary measures were proposed from three stages of design, including installation, operation and maintenance. It was recommended to improve the welding design during the design phase. In the installation stage, it was necessary to strictly control the construction quality control, attach importance to the connection of bolts and gaskets, and improve the fixing of the grounding wire. For the operation and maintenance phase, it was important to strengthen the targeted inspection of the grounding point of the cable terminal and the grounding wire, and to non-destructively test the quality of the weld during the power failure maintenance, which was an effective measure to ensure the safe operation of the grounded metal.

1. Introduction

With the development of urban power grids, the proportion of power cables in power lines is increasing, and the corresponding cable terminals are also increasing [1-2]. In order to ensure the safety of the electrical equipment and the staff under special circumstances such as breakdown or high current flow, the cable terminal is generally equipped with a grounding device, and a grounding copper block with a screw hole is welded to the tail pipe [3-4]. The wire is connected to the grounding copper block by bolts, and the other end of the grounding wire is connected to the grounding protector in the substation.

In recent years, several cable terminals in a certain urban power grid have ground faults and are closely related to grounded metals. This paper analyzes the cause of the fault in the grounding metal of the cable terminal, finds the lack of work in design and installation, and provides experience and reference for technical supervision.

2. Typical fault

Through the analysis of the ground faults of multiple cable terminals, it was found that there were three types of typical grounding metal faults, namely, the grounding copper block and the cable tail pipe weld were disconnected, the grounding copper block connecting bolt was not fastened, and the grounding wire connected to the grounding copper block was not fixed. All three cases would cause grounding metal defects, which would endanger the safe operation of the cable.

The grounding copper block and the cable tail pipe are generally connected by brazing. The weld is cracked due to stress or its own defects and gradually disengages from the tail pipe. During the gradual cracking of the weld, it is possible that the grounded copper block is completely separated from the tail pipe copper material. For example, the surface of a 110 kV cable terminal A-phase tail pipe leaves a clear trace of welding along the rectangular copper block, as shown in Figure 1(a). The grounding copper joint weld cracking chamber may also break the copper body of the tail pipe to form a hole. For example, and the terminal internal silicone oil leaks from the hole of a 110 kV cable terminal C-phase tail pipe, as shown in Figure 1(b).

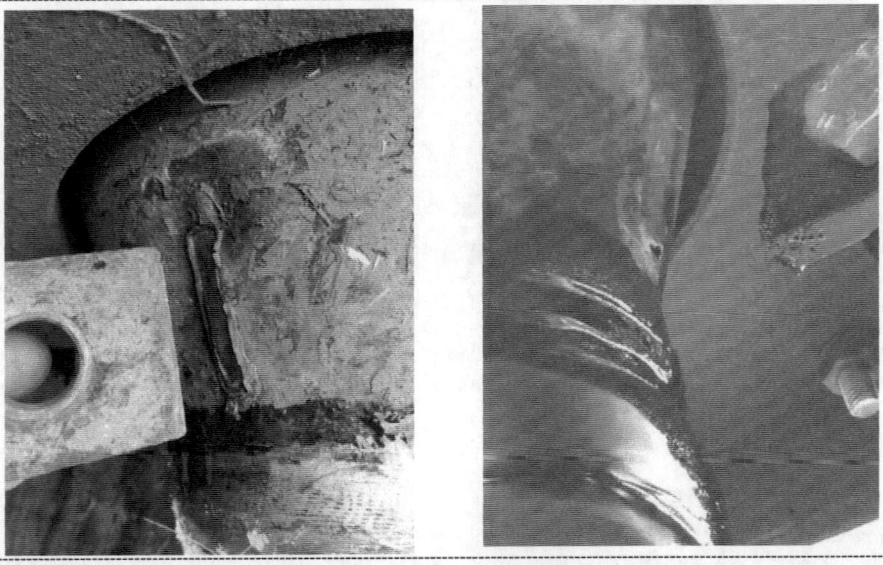

Figure 1. (a) Ground copper block and tail pipe copper are completely separated; (b) Cable terminal tail pipe copper weld zone is cracked.

The grounding copper block and the grounding wire are connected by fasteners. Because the installation process control is not in place, only a spring washer is installed on both sides of a 110 kV cable terminal A phase grounding copper block, and no flat pad is installed, as shown in Figure 2.

Figure 2. No fastened connection bolts.

Figure 3. No fixed ground wire connected to the grounding copper block.

The grounding wire connection method of the grounding copper block connection is not specified in the standard regulations and installation documents. Some grounding wires are not fixed by any measures, such as the phase B of a 110 kV cable terminal, as shown in Figure 3.

3. Analysis and discussion

The metallographic test was carried out on the grounding copper block and the weld. The bright color interval is copper matrix and the metallographic structure is normal. The gray interval is the weld structure. Inside the weld, a hole larger than 0.5 mm in size can be seen, and there are also pores of the order of 100μm.Since the copper material is not melted and there is no weld transition zone, the interface between the copper base and the weld is clear, which is a typical brazing feature, as shown in Fig. 4.

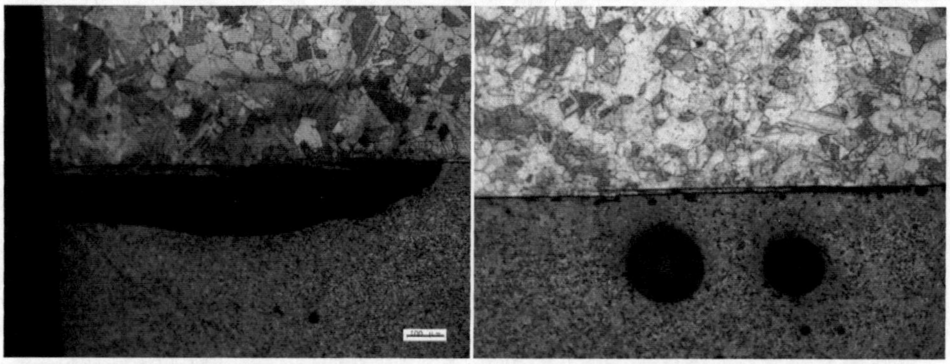

Figure 4. Metallographic structure of grounding copper block and welding.

The metallographic test shows that there is a significant weld quality defect in the brazing of the grounded copper block and the tail pipe. At the same time, according to HGJ223-1992 "Technical Regulations for Welding and Brazing of Copper and Copper Alloys", the flux should be brushed on the surface of the welding during brazing [5], and the entire welded end face of the grounded copper block should be covered with flux. From the traces of the weld crack in Figure 1, it can be seen that the grounded copper block is only partially brazed around. More importantly, the brazed weld itself has poor bearing capacity and is generally not brazed at the bearing [6]. Since the grounding copper block is brazed and welded only locally around the grounded copper block, the brazed welded joint has large holes and many pores, so its bearing capacity is poor. It is easy to detach from the tail pipe under the action of the ground wire self-weight, the external wind load installation additional torque and the corrosion stress, so that the contact failure occurs.

The connecting bolt of the grounding copper block and the grounding wire is an important node in the grounding flow guiding circuit. Flat washers and spring washers are required for bolt mounting. Gaskets and gaskets ensure good bolt contact by increasing the contact surface and preventing loose bolts. Once the bolt is missing, the bolt connection will be loose, and the contact resistance between the grounding copper block and the grounding wire will be too large. There is a safety risk in cable breakdown and high current.

Although there are no relevant standard procedures and installation specifications to clarify the way the grounding wire is fixed, it is necessary to consider the influence of the grounding wire fixing method on the service condition of the grounding copper bead weld. From the field situation, when the grounding wire is not fixed, the grounding copper block will bear the self-weight of the grounding wire and the load of the environment acting on the grounding wire, and directly promote the cracking of the welded portion of the grounding copper block.

4. Precautions

In view of the above-mentioned metal faults existing in the cable terminal, it is recommended to take measures to prevent the operation of the cable terminal from the three stages of design, installation and operation and maintenance.

(1) It is recommended to fully consider the stress of the weld between the tail pipe of the grounding copper cable end, and select the appropriate welding method and welding material.

(2) The installation requirements and standards should be strictly installed during the welding to ensure the welding quality. It is necessary to pay attention to the gasket assembly of the connecting bolts and tighten the bolts with a torque wrench. At the same time, the fixing of the grounding wire is improved to avoid additional torque to the grounded copper block weld.

(3) The operation and maintenance personnel should regularly inspect the cable terminal, and specifically detect the grounding point of the cable terminal and the grounding wire. When the power failure is overhauled, the non-destructive testing of the grounding wire welding site can be carried out, and the crack is found to be scheduled for power failure maintenance as soon as possible.

References

[1] Guo X Q, Han Y Z and Chu Q 2015 *Electric Age* **8**56. (In Chinese)
[2] Wang Z H, Xin Y C, Du C J and Li G Q 2009 *Power System Protection and Control* **2**69. (In Chinese)
[3] Song R, Li Z and Cao J R 2015 *Electronics World* **24**68. (In Chinese)
[4] Tan Y B 2018 *China High-tech* **15**105. (In Chinese)
[5] Zou J S and Han Y S 1998 *Welding Technology* **5**40. (In Chinese)
[6] Xie G J, Wang Y and Han L 2017 *Aerospace Materials & Technology* **6**79. (In Chinese)

IOP Conf. Series: Materials Science and Engineering **587** (2019) 012013 doi:10.1088/1757-899X/587/1/012013

Typical defects analysis of metal parts of substation secondary components

D K Li[1*], J Wang[1], S X Liu[2], Z Hou[2], C Qiu[3]and Y Xie[1]

[1]State Grid Hunan Electric Power Corporation Research Institute, Changsha, Hunan, 410007, China
[2]Hunan Electric Power Corporation Maintenance Company, Changsha, Hunan, 410004, China
[3]State Grid Zhuzhou Power Supply Company, Zhuzhou, Hunan, 412000, China

Corresponding author and e-mail: D K Li, 118095657@qq.com

Abstract. The selection of metal materials used in secondary components only considers the conduction of signals, ignoring the possible corrosion and fracture caused by the interaction of environmental factors such as temperature, humidity and harmful ions during the long-term service. Operational and maintenance experience shows that failures of metal parts in secondary components are emerging in endlessly due to improper selection of materials, unqualified materials and unfavourable environmental factors. In this paper, through the analysis of several secondary component failures, the typical defects of metal parts in secondary component are revealed, and some basic principles that should be followed in the selection of metal materials for secondary components are put forward, which can guide the design, acceptance and maintenance of secondary components.

1. Introduction

Secondary components are important control components in primary equipment of substation, such as relays, switching switches, auxiliary switches, travel switches, terminal row of secondary cable cores, etc. Their "healthy" service is an important guarantee that directly determines the reliable operation of primary equipment and rapid isolation of power grid faults[1].The selection of metal materials for secondary components only considers the conduction of signals, but does not consider the possible corrosion and fracture problems caused by the interaction of environmental factors such as temperature, humidity and harmful ions during the long-term service, and does not explicitly set service life. At present, the metal materials used in secondary components are various, ranging from galvanized steel, hot-dipped galvanized steel, brass and stainless steel. Several major failures of control circuit have occurred, including rejection of 500 kV circuit breaker switch and total station voltage loss of 220 kV transformer substation, due to the corrosion and fracture of metal parts in secondary component.

In this paper, through the analysis of several secondary component failures, the prominent problems existing in the metal parts of secondary components are revealed, and some basic principles that should be followed in the selection of materials for secondary components are pointed out, which can guide the design, acceptance and maintenance of secondary components.

2. Unreasonable material selection

2.1. Selection of low-grade materials

In the process of substation maintenance and operation, the failure of CA10 series switch of LW10B circuit breaker manufactured by a manufacturer has been found many times. The typical defects include damage, contact site rust and spring breakage, as shown in Figure 1, which can result in the opening and closing control failure of circuit breaker. For example, after disassembling the far and near control switch of a 500 kV circuit breaker in a substation, it is found that four of the eight rows of contacts are corroded and springs are broken, some of which are powdered. During overhaul of the 614 circuit breaker for a 220 kV transmission line in a 500 kV substation, the far and near control switch broke during operation.

Failure analysis of CA10 series switch of LW10B circuit breaker has been carried out. It was found that the metal materials for contacts, gaskets and compression springs are made of ordinary galvanized steel or 200 series stainless steel. These materials have poor corrosion resistance and are easy to rust and fracture in wet environment during long-term service[2]. The test results show that the contact spring of CA10 series switch with serious rust is made of galvanized carbon steel and its metallographic structure is fibrous tempered sorbite, as shown in figure 2. However, under the same service conditions, contact springs made of 304 stainless steel are with good surface condition, and their metallographic structure is fibrous deformed austenite and carbide[3], as shown in figure 3.

Figure 1. Defects of CA10 series far and near control switches

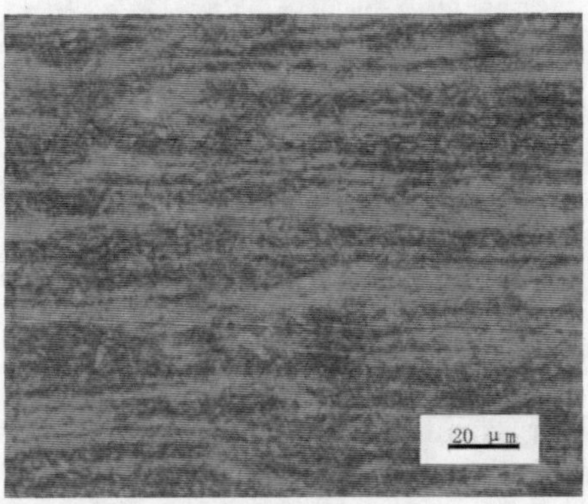

Figure 2. Metallographic structure of contact springs with serious rust

Figure 3. Metallographic structure of contact springs in good surface state

2.2. Unqualified material composition

During overhaul, it was found that the fixed bolts and nuts of a moving iron core of the opening coil of 444 circuit breaker in a 500KV substation were broken.The alloy composition of the fractured parts is analyzed, and the results are shown in Table 1.Comparing the chemical composition range of lead brass in GB/T 5231-2012 "Designation and Chemical Composition of Wrought Copper and Copper Alloys" standard, it is found that the content of Cu element in the sample is lower than the standard requirement value, while the content of Pb element is far beyond the upper limit of the standard requirement value. Therefore, material of thefixed bolts and nuts for inspection is unqualified.

Table 1. Results of chemical composition analysis of the fracture fixed bolts. Similar alloy brands in GB/T 5231-2012 are also listed for comparison.

Species	Grades	Chemical Composition/%					
		Cu	Fe	Pb	Sn	Zn	Impurity
Lead Brass	HPb62-3	60.0~63.0	0.35	2.5~3.7	-	Allowance	0.85
	HPb60-3	58.0~61.0	0.3	2.5~3.5	0.3	Allowance	0.80
	HPb58-3	57.0~59.0	0.5	2.5~3.5	0.5	Allowance	1.00
	HPb57-4	56.0~58.0	0.5	3.5~4.5	0.5	Allowance	1.20
Fixed Bolts and Nuts	Unknown	52.01	0.56	5.95	0.79	40.35	0.34

The samples were further analyzed and the results are shown in figure 4. Metallographic pictures of the polished unetched samples are shown in figure 4(a). The black island in the figure is free element Pb. Metallographic structure after etching with $FeCl_3$ hydrochloric acid solution are shown in figure4(b). The main constituent phases are equiaxed α phase and matrix β phase, and the obvious third phase element Pb is with black island shape[4].

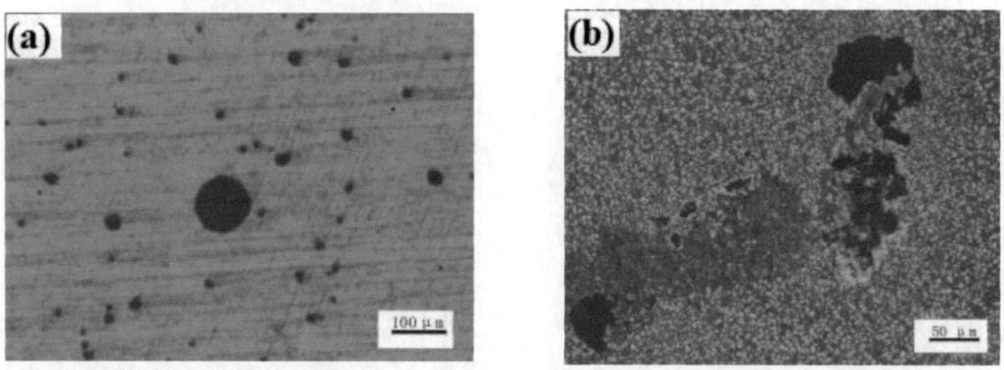

Figure 4. Microscope of the fracture fixed bolts and nuts (a) unetched after polishing (b) etched

The solid solubility of Pb in copper-zinc alloy is very small, and it mainly exists at the grain boundary of solid solution. The hardness of the element Pb is very low, it has strong lubrication and friction reduction, and it can make the alloy have very high cutting performance. However, if the content of Pb in the alloy is too high, the distribution of Pb in the grain boundary will be continuous network, leading to the weakening of the grain boundary and even the island distribution of the precipitated Pb in the matrix, as shown in figure 4. All of these will lead to the decrease of mechanical properties and the increase of brittleness of the alloy, which will easily lead to fracture[4]. Therefore, the chemical composition of the alloy does not meet the requirements of the standard, especially the excessive lead content in the alloy is the direct cause of the breakage of the fixed bolts and nuts in the moving iron core of the circuit breaker opening coil.

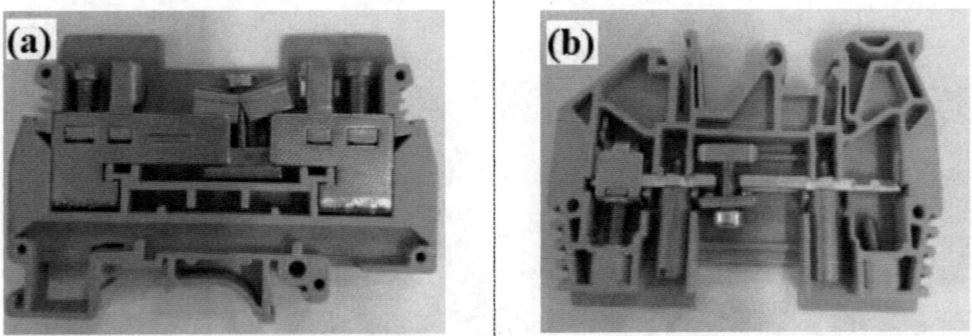

Figure 5. Terminal row (a) fractured (b) intact

During overhaul, it was found that there were a lot of fractures in one type of terminal row, as shown in figure 5(a), while other types of terminal row were intact under the same conditions, as shown in figure 5(b). Further analysis shows that the material of the broken terminal row is lead brass, and the lead content is far beyond the upper limit required by the standard, which will lead to brittleness of the terminal row. Thus, it is easy to cause cracking in the process of installation and screw hole processing.

Table 2. Chemical composition of the intact terminal row

Species	Grades	Chemical Composition/%				
		Cu	Fe	Pb	Zn	Impurity
Ordinary Brass	H66	64~68.5	0.50	0.09	Allowance	0.45
Intact Terminal Row	-	67.79	0.16	0.08	31.78	0.19

Analysis shows that there is a tin plating layer on the surface of intact terminal row. The base alloy grade is ordinary H66 brass, and its composition is shown in Table 2. The H66 brass has high content of Cu, low content of Pb and impurities, and good plasticity. So it is more suitable for terminal row material than lead brass. In addition, we also found that the terminal rows made of aluminium alloy and tin plated copper have much better durability.

3. Corrosion Caused by Environmental Factors

Terminal of the secondary cable core in terminal box of a substation is corroded, and the number barrel made of polyvinyl chloride is green. When unplugging the number barrel and unlashing the terminal, it is found that the cable core has been corroded seriously, and it almost falls off from the terminal row. The corrosion location of copper wire is just at the interface of insulating layer, but there is no obvious corrosion at a distance, as shown in figure 6. XRD and XPS analysis showed that the corrosion product of copper wire was $Cu_2Cl(OH)_3$, which was a typical corrosion product of copper and copper alloy in marine atmospheric environment.

Further analysis shows that Cl^- can be dissolved from the number barrel and dominates the whole corrosion process. When the external humidity increases, thin liquid film is formed on the surface of copper wire, copper will react with oxygen in air to form Cu_2O film, and the corrosion rate decreases. In the meantime, a thin liquid film will be formed around the number barrel, and the chlorine elements

IOP Conf. Series: Materials Science and Engineering **587** (2019) 012013 doi:10.1088/1757-899X/587/1/012013

in it will dissolve in the form of chloride ions, which will increase the concentration of Cl^- in the surrounding environment. Then, Cl^- gradually invades into Cu_2O film and replaces O element in it to form soluble $CuCl_2^-$, which is further oxidized to soluble $CuCl_2$ under the action of oxygen. Whereafter, $Cu_2Cl(OH)_3$are formed by coordination reaction and redeposition process between Cu^{2+}, Cl^- and OH^-. The whole process can be expressed as follows:

$$Cu_2O + Cl^- + 2H_2O + 1/2O_2 \rightarrow Cu_2Cl(OH)_3 + OH^- \quad (1)$$

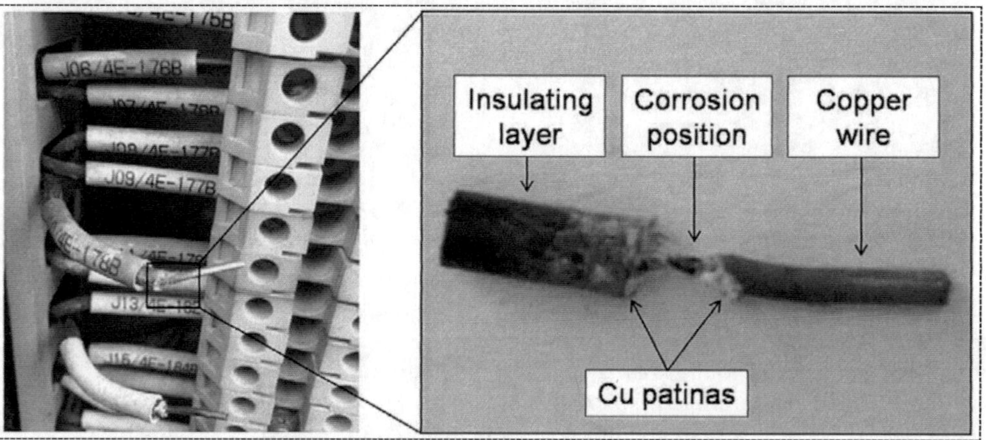

Figure 6. The damaged copper wires in outdoor terminal boxes[5]

From the above discussion, it can be seen that Cl^- dissolved from the number barrel and higher air humidity are the main environmental factors accelerating the corrosion process of copper conductors. In order to avoid the corrosion of copper conductors, it is suggested to replace the dehumidifier with independent control and evaluate the dehumidification effect regularly and the number barrel should be replaced with the material which does not dissolve the harmful ions such as Cl^- andSO_4^{2-}.

The above cases show that the influence of environment, including temperature and humidity, harmful ions, and potential galvanic corrosion or crevice corrosion, should be fully taken into account when selecting metal materials for secondary components.

4. Conclusions

Due to improper selection of materials, unqualified materials and unfavorable environmental factors in the long-term service process, failures of the metal parts of secondary components emerge in endlessly. In order to ensure the reliability of secondary components in service and to reduce or even eliminate the control circuit faults caused by the metal parts failures of secondary components, it is suggested that the following work be carried out:

(1) Technical standards such as guidelines for selection of metal materials and quality acceptance criteria for secondary components should be formulated as soon as possible, and the quality acceptance of secondary components into the power grid should be strengthened.

(2) In the design stage, material with excellent comprehensive properties such as aluminum alloy, nickel-plated brass and tin-plated copper should be preferred, while material with insufficient corrosion resistance such as galvanized steel, 200 series stainless steel and poor mechanical properties such as lead brass should be avoided.

(3) In order to ensure the reliability of secondary components in service life, effects of temperature, humidity, harmful ions and other environmental factors on the metal parts of secondary components should be studied in detail. Meanwhile, it is necessary to establish a method for evaluating the service life of different metal materials under various environmental conditions.

IOP Conf. Series: Materials Science and Engineering **587** (2019) 012013 doi:10.1088/1757-899X/587/1/012013

References

[1] Liu J X 2017 *Science and Technology in China* 06: 76-80. (In Chinese)
[2] Yang H Y, Ding G Q, Huang G Q, et al. 2017 *Corrosion & Protection* 05: 369-372. (In Chinese)
[3] Hou X Z, Zheng W J, Song Z G, et al. 2013 *Journal of Iron and Steel Research* 07: 53-57. (In Chinese)
[4] Zhang M, Wu C Q, Li Y 2009 *Material Research & Application* 03: 200-204. (In Chinese)
[5] Wu T Q, Zhou Z F, Xu S, et al. 2019 *Engineering Failure Analysis* 98: 83-94.

IOP Conf. Series: Materials Science and Engineering **587** (2019) 012014 doi:10.1088/1757-899X/587/1/012014

Research on rapid qualitative detection of lead in pyrotechnic powder used for fireworks and firecrackers

Junyi Wu[1]

[1] Beihai Customs Comprehensive Technology Service Center, Beihai, Guangxi, China

*Corresponding author and e-mail: johnny2000@vip.qq.com

Abstract. This study discloses a method for qualitatively quick detecting the leading element for fireworks and firecrackers based on energy dispersive X-ray fluorescence spectrometer (EDXRF), including the following steps: preparation of samples, establishment of detection methods, determination of the characteristic line fluorescence intensity values of Sb element in samples. The method of the study has the advantages like that: (1) the method is simple to operate, and the method can be repeatedly called for testing. Only one new test method needs to be built before the sample test. After the method is established, the test can be repeated at different times without re-establishing the test method for each test. After the first establishment of the new test method, the entire test process only includes three steps: sample preparation, sample loading into the sample cup and on-board testing. (2) The detection period is extremely short. After the sample is prepared, the entire measurement process takes only about 2 minutes. (3) Labor intensity is very low and the requirements for operators are not high. (4) The method has good stability, good repeatability and high credibility.

1. Introduction

According to the literature report, the qualitative detection method of lead in pyrotechnics for fireworks and firecrackers is based on traditional chemical analysis methods. The currently used method standard is "Rules for the inspection of export fireworks and firecracker-Part 4:Safety performance test" (SN/T 0306.4-2006) Such standard methods have many disadvantages such as cumbersome operation steps, low detection efficiency, high use of reagents, large environmental pollution, and large influence of human factors such as coloring reaction determination endpoints.In the electronic product testing standard, there is a standard method for analyzing heavy metals in electronic products. And the commonly used method standard is "Rapid screening of lead, cadmium, chromium, lead and bromine of regulated substances in electrical and electronic equipment-X-ray fluorescence spectrometry (GB/Z 21277-2007). The analytical instrument used in this standard method is a wavelength dispersive X-ray fluorescence spectrometer, the basic principle of which is completely different from the principle described in the present invention.

The methods currently developed by energy dispersive X-ray fluorescence spectrometers (EDXRF) are mostly used for nondestructive qualitative analysis of samples. For semi-quantitative and quantitative elemental detection of solid samples, most samples are directly determined by powder tableting and melting. Because pyrotechnics for fireworks and firecrackers are flammable and explosive, it is impossible to use the powder tableting method and the melting method for sample processing. So far, there has not been a publicly reported on a method for rapidly and qualitatively detecting lead element in pyrotechnics for fireworks and firecrackers based on energy dispersive X-ray fluorescence spectroscopy.

2. Theory

As we know, the fluorescence intensity of lead element of the sample powder depends on the content of lead in pyrotechnic based on energy dispersive X-ray fluorescence spectroscopy.After the sample is excited by X-rays, different elements in the sample emit different characteristic lines, which are fingerprint information of identifying the target elements in the sample.According to the characteristics of the chemical composition of lead used for fireworks and firecrackers, a special mathematical model is established. Optimize the various factors which directly affect the results of the measurement, including: the type of method used to establish the analytical method and the voltage of the energy dispersive X-ray fluorescence spectrometer, the current, filter, peak spectrum observation line selection, analysis time, count rate, gas environment, energy range and the thickness of the sample in the sample cup and ect. Based on the assumption that the content of lead in the powder sample is positively correlated with the fluorescence intensity of the lead characteristic line and the ratio of the content of the lead element to the fluorescence intensity of the characteristic line of the lead element is a fluctuation within a certain range, it can conclude that the lead content in the sample can be detected qualitatively by detecting the fluorescence intensity of the characteristic line of the lead element in the sample.

3. Experimental section

3.1. Instrument and apparatus

Oven with accuracy to ±2°C. Analytical balance with accuracy to 0.1 mg. energy dispersive X-ray fluorescence spectrometer (EDXRF): United States Thermo Fisher (former Thermo Electron Corporation) Company QUANT'X series.

3.2. Operation step

(1) 5 to 10 g of the 40-100 mesh sieve sample powder is thoroughly mixed, placed in an oven, dried, placed in a desiccator and cooled to room temperature, and ready to be used.

(2) Weigh the sample of about 1 g, make sure the thickness of the powder sample in the sample cup is ≥3mm.

(3) Gently tamper the sample cup 3 times on the hard ground and put the cup in the testing tank.

(4) Set the parameters of the EDXRF instrument as shown in Table 1.

Table 1. Parameters of the EDXRF instrument

Filter	Thick Pd
Collimator	8.8mm
Voltage	28v
Electric current	Auto
Analysis time	30s
Count rate	Medium
Atmosphere	Air
Matrix effects	Not considered
Energy range	0～40kev
Analysis technique	Intensity correction
sample thickness	≥3mm

(4) Sample determination: determine the fluorescence intensity of the target element of the sample under the best analysis condition and read the values of it.

IOP Conf. Series: Materials Science and Engineering **587** (2019) 012014 doi:10.1088/1757-899X/587/1/012014

4. Results and Discussion

4.1. Sample size and particle size

In the method, 5 to 10 g of the 40-100 mesh sieve sample powder is thoroughly mixed, placed in an oven, dried, placed in a desiccator and cooled to room temperature, and ready to be used. The reason why the particle size of the sample is set to 5 ~ 10g is that in the actual production process, the quality of the pyrotechnics for fireworks and firecrackers is uneven and the density of the pyrotechnic composition is high, if the sample size is too small, the sample would not be representative and would be difficult to meet the requirements of the sample thickness in the sample cup which is required over 3mm thickness, and it will directly affect the accuracy of the test results. If the sample size is too large, it will affect the efficiency of the sample preparation.

There are two main reasons why the sample must be passed through a 40-100 mesh sieve: Firstly, The energy dispersive X-ray fluorescence spectrometer analyzes the surface of the sample to get the fluorescence intensity of the characteristic line of lead element, if the sample with uneven particle size is likely to have a large particle size effect which would seriously affect the accuracy of the test results. So it must be sure to make the particle size of the sieved sample not to be too big to avoid increasing unevenness of particle size of the sample. A large amount of experimental data indicates that the particle size of the sieved sample is less than 40 mesh would cause little particle size effects. Secondly, if the powder sample passes through a sieve of more than 100 mesh, the particle size will become very small, and which will not only affect the screening efficiency of the sample but also increase the dust concentration in the environment due to the too small powder particles after the screening. It is also a certain health hazard to the sample preparation personnel. Another important reason is that the pyrotechnic sample powder with a particle size of less than 100 mesh has flammability and is easily ignited in the air.

4.2. Rules for Judgment

Different countries have different regulations on the use of prohibited substances in fireworks and firecrackers. For example, the relevant standards in the American Pyrotechnics Association stipulate that substances within 0.25% by mass of pyrotechnics are recognized as impurities. China's national standard "Safety and Quality of Fireworks and Firecracker" (GB 10631-2013) stipulates that the substance within the concentration of 0.1% by mass of pyrotechnics is recognized as an impurity, and the Netherlands found that the lead content of a certain kind of fireworks from China exceeds 120mg/kg in the results of an imported fireworks sampling test and Announced it. In combination with the actual situation of pyrotechnics for fireworks and firecrackers in China and the characteristics of energy dispersive X-ray fluorescence spectrometers, in order to effectively solve the practical problems of qualitative detection of heavy metals in fireworks and firecrackers, this method is based on the relevant content of National Standard (GB 10631-2013) to define the effective detection limit of lead element is 1000 mg/kg (0.1% by mass).

4.3. Advantages

The method is based on the energy dispersive X-ray fluorescence spectroscopy technology for quickly qualitatively detecting the lead element for fireworks and firecrackers, and the advantages thereof are as follows: (1) The method is simple to operate, and the method can be repeatedly called for testing. Only one new test method needs to be built before the sample test, and after the method is established, the test can be repeated at different times without re-establishing the test method for each test. After the establishment of the new test method, the entire test process only includes three steps: sample preparation, sample loading into the sample cup and on-board testing. (2) The detection period of the method is extremely short, and after the preparation of the sample, the entire measurement process only takes about 2 minutes. (3) The method has low labor intensity and is not demanding to the operator. (4) The accuracy is good, the precision is high, and the false positive rate is low.

4.4. Method for Validation Test

Because the standard of pyrotechnics with a certain amount of lead content can not be found in the market, and the physical form of black powder is similar to that of pyrotechnics, the reference material for the different lead content of black powder as the matrix configured with the standard material of lead oxide can be tested as the samples. By comparing the correspondence between the lead content of different pyrotechnic reference materials and their corresponding characteristic fluorescence intensity values, the general correspondence between the lead content in the pyrotechnic composition and its corresponding characteristic fluorescence intensity would be inferred. The numerical relationship between the fluorescence intensity value and the content value of the lead element in the samples can be seen in Table 2.

Table 2. The numerical relationship between the fluorescence intensity value and the content value of the lead element

P content (mg/kg)	0	200	400	600	800	1000	1200	1400	1600
P Fluorescence intensity values(cps/mA)	0	9	17	26	36	45	57	59	73

It can be seen from Table 2 that: When the content of lead is in the range of 0 to 1600 mg/kg, the fluorescence intensity value of the characteristic line of lead element increases with the increase of lead oxide content, which is positively correlated. And when the content of lead is in the range of 0 to 1200 mg/kg, it is substantially proportional. In particular, when the content of lead is 1400 mg/kg, the fluorescence intensity value of the lead element characteristic line (59 cps/mA) is only 2 cps/mA higher than the fluorescence intensity value at 1200 mg/kg content (57 cps/mA). Obviously, the increase of the fluorescence intensity value is not proportional to the lead content. The main reason is that with the increase of lead content in the sample, each element in the sample has an increasingly obvious matrix effects on the lead element and this matrix effects will increase the fluorescence intensity value of the lead element characteristic line randomly, sometimes the increasing amount will reduce or even be negative growth. However, when the content of lead is ≥0.1%, the fluorescence intensity value of the characteristic line of the lead element is always ≥45 cps/mA. Therefore, it can be used as the basis for detecting whether the sample contains the lead element content above 0.1% or not that the fluorescence intensity value of the characteristic line of lead element is above 45 cps/mA.

5. Conclusions

This method discloses a method for quickly qualitatively detecting the lead element for fireworks and firecrackers based on EDXRF with high accuracy, good repeatability, simple operation and high efficiency. It can effectively meet the rapid detection of lead in pyrotechnic samples by manufacturers, regulatory authorities and third-party laboratories.

References

[1] Leif, H. C.; Allan, A. Determination of sulfur and heavy metals in crude oil and petroleum products by energy-dispersive x-ray fluorescence spectrometry and fundamental parameter approach.Anal. Chem.,53(12), pp.1788-1792,1981.

[2] Fei, H; Pierre, J. V. E. General approach for quantitative energy dispersive x-ray fluorescence analysis based on fundamental parameters. Anal. Chem.,63(20), pp.2237-2244,1991.

[3] Standard test method for determination of lead in paint layers and similar coatings or in substrates and homogenous materials by energy dispersive x-ray fluorescence spectrometry using multiple monochromatic excitation beams, ASTMF2853, American Society for Testing and Materials Publications, 2010.

[4] Standard test method for analysis of uranium and thorium in soils by energy dispersive x-ray fluorescence spectroscopy, ASTM C1255, American Society for Testing and Materials Publications, 2011.

[5] Duan Tiyu, Li Suqing, et al. Jewellery, Determination of precious metal content - Method using

x-ray fluorescence spectrometry, China National Recommended Standard: GB/T 18043-2013, China Standard Press: Beijing, 2014.

[6] Leoni. L, Saitta. M, X-ray fluorescence analysis of powder pellets utilizing a small quantity of material, X-ray Spectrom, 3, pp.74-77,1974.

Research on rapid semi-quantitative determination of aluminum in pyrotechnics used for fireworks and firecrackers

Junyi Wu[1]

[1]Beihai Customs Comprehensive Technology Service Center, Beihai, Guangxi, China

*Corresponding author and e-mail: johnny2000@vip.qq.com

Abstract. This study discloses a method for rapid semi-quantitative determination of aluminum in pyrotechnic powder for fireworks and firecrackers based on energy dispersive X-ray fluorescence spectrometer (EDXRF), including the following steps: preparation of samples, establishment of testing methods, determination of the characteristic line fluorescence intensity values of Al element in samples , and according to the measured fluorescence intensity values, the aluminum content in the pyrotechnic powder samples can be semi-quantitatively determined. The method of the study has the advantages like that: (1) the method is simple to operate, and it can be repeatedly called for testing. Only one new test method needs to be built before the sample test. After the method is established, the test can be repeated at different times without re-establishing the test method for each test. After the first establishment of the new test method, the entire test process only includes three steps: sample preparation, sample loading into the sample cup and on-board testing. (2) The testing time is extremely short. After the sample is prepared, the entire measurement process takes only about 2 minutes. (3) Labor intensity is very low and the requirements for operators are not very strict. (4) The method has good stability, good repeatability and high credibility.

1. Introduction

In the prior art, no method for quickly determining the aluminum content in pyrotechnics used for fireworks and firecrackers based on energy dispersive X-ray fluorescence spectroscopy has been found, and only the method for determining aluminum content in pyrotechnics for fireworks and firecrackers has been found in the National standard of "Determination of aluminum content in pyrotechnics used for fireworks and firecrackers" (GB/T 20615-2006) and determination method of active aluminum content in aluminum powder in the National standard of " Chemical analysis method of Aluminium powder ,gas volume method for the determination of active aluminum" (GB 3169.1-1982), such methods are both based on traditional chemical analysis to quantitatively analyze the aluminum content in the sample. Taking the method of "Measurement of aluminum content in pyrotechnics for fireworks and firecrackers " (GB/T 20615-2006) as an example, the basic principle of this standard method: After proper pretreatment, the sample is firstly dissolved with dilute nitric acid and filtered, and then the aluminum powder in the filter residue is dissolved with sulfuric acid, and filtered, and then the filtrate is collected, the pH of the filtrate is adjusted to some value, and the aluminum ion in the test solution is separated by adding a precipitant. Using EDTA complexometric titration method under the condition of pH10.0 with PAN as the indicator solution, one sample solution was added with appropriate amount of ammonium fluoride to mask the aluminum ions in the test solution, and the other test solution was directly added with EDTA, and then using copper sulfate as standard titration solution to titrate the sample solution from yellow to blue as the end point, and according to the volume difference of the copper sulfate standard titration solution consumed by the two sample solutions to calculate the aluminum content of the sample.

The method described in this standard has the following deficiencies: (1) The detection period is long. It will takes a skilled technician two working days to complete a test. In addition, it is easy to introduce uncertainty due to insufficient proficiency of the tester during the specific test process. (2) The operation steps are cumbersome. The sample is washed several times with absolute ethanol and acetone. After being dissolved in dilute nitric acid, it is filtered, transferred, and collected. Then the sulfuric acid is used to fully dissolve the filter residue, and then subjected to filtration, washing and pH adjustment. The precipitant is used to separate the aluminum ions, and finally the amount of aluminum in the solution is determined by a differential method. (3) The method requires high requirements for the tester. Many steps in the operation steps are easy to introduce uncertainties such as washing, transfer, dissolution, filtration, sedimentation, enrichment, titration, etc. Each tester must be extra careful and meticulous. Otherwise, it is very easy to introduce artificial uncertainty.

The methods currently developed by energy dispersive X-ray fluorescence spectroscopy are mostly used for non-destructive qualitative analysis of samples. For semi-quantitative and quantitative elemental detection of solid samples, most samples are directly determined by powder tableting or melting methods, such as: "Determining the content of precious metals by X-ray fluorescence spectrometry" (GB/T 18043-2008) using non-destructive Detection method, "Alumina chemical analysis method and physical property measurement method Part 30 X-ray fluorescence spectrometry determination of element content" (GB/T6609.30-2009) using the melting method, "EDXRF method for direct determination of W-Fe- Ni-Co Alloy Mixture Components ("Nuclear Electronics and Detection Technology" Issue 5,2007) uses the tabletting method, "X-ray fluorescence spectrometry for rapid determination the contents of potassium, sodium, calcium and magnesium in potassium chloride products". (Analytical Instruments, Issue 6, 2013) using the tableting method, "Rapid screening X-ray fluorescence spectrometry for lead, mercury, chromium, cadmium and bromine in electronic and electrical products" (GB/Z 21277) -2007) uses the tableting method or the melting method. Because pyrotechnics for fireworks and firecrackers are flammable and explosive, it is impossible to tablet or melt the pyrotechnic powder for the sample preparation. So far, there has been no published literature report on the method of rapid semi-quantitative determination of aluminum content in pyrotechnics for fireworks and firecrackers based on energy dispersive X-ray fluorescence spectrometry.

2. Theory

After the pyrotechnic sample is excited by X-rays, different elements in the sample emit different characteristic X-rays. These characteristic lines are fingerprint information that identifies the target element in the sample. By measuring the characteristic X-ray fluorescence intensity of the target element in the sample, it is possible to semi-quantitatively analyze the amount of aluminum in the unknown pyrotechnic sample. The method directly uses the pyrotechnic powder sample of the fireworks and firecrackers to establish a specific analysis method, and semi-quantitatively analyzes the aluminum content in the sample according to the X-ray fluorescence intensity value of the characteristic line of the aluminum element.

3. Experimental section

3.1. Instrument and apparatus

Oven with accuracy to ±2°C. Analytical balance with accuracy to 0.1 mg. energy dispersive X-ray fluorescence spectrometer (EDXRF): United States Thermo Fisher (former Thermo Electron Corporation) Company QUANT'X series.

3.2. Operation step

(1) 10 to 30 g of the 40-100 mesh sieve sample powder is thoroughly mixed, placed in an oven, dried, placed in a desiccator and cooled to room temperature, and ready to be used.

(2) Weigh the sample of about 2 g, make sure the thickness of the powder sample in the sample cup is ≥3mm.

(3) Gently tamper the sample cup 3 times on the hard ground and put the cup in the testing tank.

(4) Set the parameters of the EDXRF instrument as shown in Table 1.

Table 1. Parameters of the EDXRF instrument

Filter	No filter
Collimator	8.8mm
Voltage	20v
Electric current	Auto
Analysis time	30s
Count rate	Medium
Atmosphere	Vacuum
Matrix effects	Not considered
Energy range	0～40kev
Analysis technique	Intensity correction
sample thickness	≥3mm

(4) Sample determination: determine the fluorescence intensity of the target element of the sample under the best analysis condition and read the values of it.

(5) Calculate the amount of aluminum in the sample by calculating the fluorescence intensity value according to the following formula:

$$\omega = \frac{F}{C}$$

Wherein, ω indicats the mass percentage of aluminum in the sample, the unit is %; C is the fluorescence intensity value of the corresponding characteristic line when the mass percentage of aluminum element in the sample is 1%, and the unit is cps/mA; F is the fluorescence intensity value of the characteristic line of the aluminum element measured by the energy dispersive X-ray fluorescence spectrometer in units of cps/mA.

4. Results and Discussion

4.1. Sample size and particle size

In the method, 10 to 30 g of the 40-100 mesh sieve sample powder is thoroughly mixed, placed in an oven, dried, placed in a desiccator and cooled to room temperature, and ready to be used. The reason why the particle size of the sample is set to 10 ~ 30g is that in the actual production process, the quality of the pyrotechnic powder for fireworks and firecrackers is uneven and the density of the aluminum powder is high, if the sample size is too small, the sample would not be representative and would be difficult to meet the requirements of the sample thickness in the sample cup which is required over 3mm thickness, and it will directly affect the accuracy of the test results. If the sample size is too large, it will affect the efficiency of the sample preparation. There are two main reasons why the sample must be passed through a 40-100 mesh sieve: Firstly, the energy dispersive X-ray fluorescence spectrometer analyzes the surface of the sample to get the fluorescence intensity of the characteristic line of aluminum element, if the sample with uneven particle size is likely to have a large particle size effect which would seriously affect the accuracy of the test results. Therefore it must be sure to make the particle size of the sieved sample not to be too big to avoid increasing unevenness of particle size of the sample. A large amount of experimental data indicates that the particle size of the sieved sample is less than 40 mesh would cause little particle size effects. Secondly, if the pyrotechnic powder sample passes through a sieve of more than 100 mesh, the particle size will become very small, and which will not only affect the screening efficiency of the sample but also increase the dust concentration in the environment due to the too small aluminum powder particles after the screening. It is also a certain health hazard to the sample preparation personnel. Another important reason is that the pyrotechnic powder with a particle size of less than 100 mesh has flammability and is easily ignited in the air.

4.2. Advantages

A method for rapid and semi-quantitative determination of aluminum content in pyrotechnic samples of fireworks and firecrackers based on energy dispersive X-ray fluorescence spectrometry, the advantages of which are as follows: (1) The method is simple to operate, and the method can be repeatedly called for testing. Only one new test method needs to be built before the sample test, and after the method is established, the test can be repeated at different times without re-establishing the test method for each test. After the establishment of the new test method, the entire test process only includes three steps: sample preparation, sample loading into the sample cup and on-board testing. (2) The analysis time of the method is extremely short, and after the preparation of the sample, the entire measurement process only takes about 1 minute. (3) The method has low labor intensity and is not demanding to the operator. (4) The accuracy is good, the precision is high, and the false positive rate is low.

4.3. Method validation test

Because the standard of pyrotechnics with a certain amount of aluminum content can not be found in the market, and the physical form of black powder is similar to that of pyrotechnics, the reference material for the different aluminum content of black powder as the matrix configured with the standard material of aluminum powder can be tested as the samples. By comparing the correspondence between the aluminum content of different pyrotechnic reference materials and their corresponding characteristic fluorescence intensity values, the general correspondence between the aluminum content in the pyrotechnic composition and its corresponding characteristic fluorescence intensity would be inferred. The numerical relationship between the fluorescence intensity value and the content value of the aluminum element in the samples can be seen in Table 2.

Table 2. The numerical relationship between the fluorescence intensity value and the content value of the aluminum element

Sample No.	1	2	3	4	5	6
Al content (%)	0	10	30	50	80	99.9
Al Fluorescence intensity values(cps/mA)	0	578	1725	2941	4925	6040
Ratio	0	57.8	57.5	58.8	61.6	60.5

It can be seen from Table 2 that: Firstly, When the sample does not contain aluminum, the fluorescence intensity value of the characteristic line of the aluminum element in the corresponding method is also zero. Secondly, Observing the point where the aluminum content differs greatly, the fluorescence intensity value of the corresponding characteristic line is enhanced with the increase of aluminum content, which is positively correlated, but not strictly proportional. The main reason is that the matrix effects of each element in the sample on the aluminum element is more obvious due to the increase of the content of aluminum, and the direct effects of these matrix effects will increase the corresponding difference in the fluorescence intensity value of the characteristic line of the aluminum element. Thirdly, the mass percentage (%) of aluminum in the sample is positively correlated with the ratio of the corresponding aluminum element characteristic line fluorescence intensity value (cps/mA), and the ratio is within a range of 1:60±10% (specific value 1:57.5 ~ 61.6).

4.4. Method repeatability test

The purpose of the method repeatability test is to further confirm the corresponding characteristic fluorescence intensity values of aluminum elements when aluminum has a mass percentage of 10% in different types of pyrotechnic agents. The samples of pyrotechnic reference materials with an aluminum content of 10% were determined by using different types of pyrotechnic agents as substrates. The specific test results are shown in Table 3.

IOP Conf. Series: Materials Science and Engineering **587** (2019) 012015 doi:10.1088/1757-899X/587/1/012015

Table 3. Fluorescence intensity values of characteristic lines of aluminum in different pyrotechnic reference materials with an aluminum content of 10%

Sample No.	1	2	3	4	5
Pyrotechnic effects	Red	Blue	Yellow	Green	White
Al Fluorescence intensity values(cps/mA)	578	610	594	585	613
Ratio	57.8	61.0	59.4	58.5	61.3

It can be seen from the test data of Table 3 that when the content of aluminum element in the sample of different pyrotechnic composition is 10%, the fluorescence intensity value of the corresponding aluminum element characteristic line ranges from 578 to 613 cps/mA. The ratio of the content of aluminum in the pyrotechnic composition to the corresponding fluorescence intensity value of the aluminum element is also positively correlated and fluctuates within a certain range (specific values are 1:57.8 to 61.3). Since it is a semi-quantitative test, this proportional coefficient can be approximated by an integer multiple of 1:57 to 1:62.

5. Conclusions

This article discloses a method for quickly and semi-quantitatively detecting aluminum content in pyrotechnic blind samples of fireworks and firecrackers based on energy dispersive X-ray fluorescence spectrometer (EDXRF), and the method has the advantages of simple operation, short detection period, good stability, good repeatability and high credibility.

References

[1] Leif, H. C.; Allan, A. Determination of sulfur and heavy metals in crude oil and petroleum products by energy-dispersive x-ray fluorescence spectrometry and fundamental parameter approach.Anal. Chem.,53(12),pp.1788-1792,1981.

[2] Duan Tiyu, Li Suqing, et al. Jewellery,Determination of precious metal content - Method using x-ray fluorescence spectrometry,China National Recommended Standard: GB/T 18043-2013, China Standard Press: Beijing, 2014.

[3] Leoni. L,Saitta. M, X-ray fluorescence analysis of powder pellets utilizing a small quantity of material, X-ray Spectrom, 3,pp.74-77,1974.

[4] Rose W.I.,Bornhorst T.J.,Sivonen S.J., Rapid high-quality major and trace element analysis of powdered rock by x-ray fluorescence spectrometry, X-ray Spectrom, 15,pp.55-60,1986.

[5] Gy, Pierre M,The analytical and economic importance of correctness in sampling,Anal.Chim.Acta, 190,pp.13-23,1986.

[6] ZHOU Tong-hui ,WANG Er-kang,LU Wan-zhen ect. Handbook in analytical chemistry(second edition), basic knowledge of and safety knowledge(the first volume) , China Chemical industry press:pp.568-580,1997.

IOP Conf. Series: Materials Science and Engineering **587** (2019) 012016 doi:10.1088/1757-899X/587/1/012016

Synthesis of Ag doped SiO2-TiO2 aerogels with nano-sized microcrystalline anatase structure through IL control

Hongyi Dai*, Zhao Jun, Yuxin Yin, Guolin Shao, Chunling Yu

School of Light Industry and Chemical Engineering, Dalian Polytechnic University, Dalian 116034, China

dai_hy@163.com, kittyufo@163.com, yinyx@dlpu.edu.cn, shaoguolin@dlpu.edu.cn, ycl223@163.com

*Corresponding Author E-mail: dai_hy@163.com

Abstract. In this paper, Ag doped SiO2-TiO2 aerogels with nano-sized microcrystalline anatase structure through a sol-gel way, ionic liquid(IL), 1 – butyl- 3 - methylimidazole bromide ([bmim]Br), is used as a structure control agent. The catalysts were characterized by SEM, XRD and nitrogen adsorption-desorption isotherms (BET analyses). The photocatalytic activities were investigated by degration of active brilliant blue under UV-light. It was demonstrated that IL facilitate the formation of mesoporous and anatase microcrystalline structure at around ambient condition, without calcination. Both at sol-step and gel-step, the concentration of IL aqueous solution prominently effected on the structure properties of catalysts. Surface area, crystalline grain dimension and mecroporous structure all played important roles in the degradation reactions. The catalyst with 1/10 Si/Ti ratio, 385.2 m2g-1surface area and 6.2nm crystalline grian size, showed excellent photocatalytic degradation rate up to 99.7%.

1. Introduction

TiO2, regarded as the most efficient environmental semiconductor photocatalyst has been widely studied over past sixty years. The principle of photocatalysis is absorption of photons with energy larger than the band gap of TiO2, electrons are excited from the valence band to the conduction band, creating electron-hole pairs. These charge carriers migrate to the surface and react with the chemicals adsorbed on the surface to decompose chemicals. The photocatalytic activity of a semiconductor is largely controlled by (i) the light absorption properties, e.g., light absorption spectrum and coefficient, (ii) reduction and oxidation rates on the surface by the electron and hole, (iii) and the electron-hole recombination rate. A large surface area with a constant surface density of adsorbents leads to faster surface photocatalytic reaction rates. In this sense, the larger the specific surface area, the higher the photocatalytic activity is. On the other hand, the surface is a defective site; therefore, the larger the surface area, the faster the recombination. The higher the crystallinity, the fewer the bulk defects, and the higher the photocatalytic activity is. Hightemperature treatment usually improves the crystallinity of TiO2 nanomaterials, which in turn can induce the aggregation of small nanoparticles and decrease the surface area. Optimal conditions are sought by taking these considerations into account and may vary from case to case[1-2].

Silica - titania is one of the important binary oxide materials and has been widely used for academic purpose and industry [3-5]. It has attracted widely interest due to their potential applications as the catalyst in various reactions from fine chemistry to photocatalytic removal of the water pollutant. The properties of silica - titania materials depend on the structure, composition, and homogeneity of the prepared material [5].

Aerogels have been receiving a considerable attention because of their unique properties, i.e., very low density, extremely high surface area, high porosity, low thermal conductivity and low dielectric permittivity [1-5]. TiO2 has been used to treat rarious industrial wastewater owing to its excellent photocatalysis activity. However TiO2 aerogels are restricted because of its poor network structure. SiO2-TiO2 aerogels have higher surface area, stronger network structure, higher absorbing capacity and better photocatalytic properties than TiO2 aerogels [6].

SiO2-TiO2 aerogels are of immense importance in various applications such as adsorption, sensing, and catalysis.

Nowadays, ionic liquids (IL) have been used as effective solvents or co-solvents or template to synthesize aerogels by ambient pressure drying instead of using supercritical drying because of their excellent thermal stability, highly conductivity, wide electrochemical widow, negligible vapor tension and special solubility [7-9]. Sheng Dai et al. [10] have successfully used IL as effective solvents to synthesize silica aerogels with high surface area (720 m2g-1), large pore volume (1.4 cm3g-1) and stable aerogel network structures without the need for supercritical drying processes. Kyesang Yoo et al. [11] synthesized anatase-containing nanostructured TiO2 particles with surface area of 273 m2g-1 using ionic liquid 1-butyl-3-methylimidazolium

hexafluorophosphate as an effective additional solvent by the sol–gel method at low temperature. My team has successfully prepared SiO2-TiO2 aerogels with high surface area and good photocatalysis activity using IL by ambient pressure drying[12,13]. Usually, SiO2-TiO2 aerogels don't have anatase crystal structure before calcinations. It is well known that crystal types of include anatase, brookite and rutile, among which the anatase possesses the best photocatalyst activity. Traditionally, anatase TiO2 is obtained by calcinations which can reduce surface area of aerogels. So, it is important to obtain anatase TiO2 avoiding calcinations.

In this paper, Ag doped SiO2-TiO2 aerogels were synthesized at ambient pressure by using IL, 1-butyl-3-methylimidazolium bromide ([bmim]Br) as cosolvent and aging agent. In the two step sol-gel process, tetraethylorthosilicate (TEOS) and tetrabutyl titanate (TBT) were used as precursors and glacial acetic acid and urea were used as catalysts. Acetic acid glacial as a kind of weak acid could decrease the hydrolytic rate of precursors and urea as a kind of weak base could control the rate of polycondensation reaction so that TEOS and TBT would precede hydrolysis and polycondensation simultaneously. The influence of the SiO2/TiO2 molar ratios, the IL/(SiO2+TiO2) molar ratio ratios, the different dosage of Ag on aerogels was investigated and the aerogels were characterized by nitrogen adsorption analysis, scanning electron microscopy (SEM) and X-ray diffraction (XRD) . The photocatalytic properties of the aerogels were also evaluated by the degradation of reactive brilliant blue in aqueous solution under UV-light.

2. Experimental

2.1. Preparation of aerogels

Tetraethyl orthosilicate (TEOS) and butyl titanate (TBT) as precursors were obtained from Dongtai Tianjin Kemiou Chemical Reagent Co. and used as received. Acetic acid glacial and urea as catalysts was produced by Tianjin Dongzheng fine chemical reagent factory. Miconazole and butyl bromide used to synthesize [bmim]Br were bought from Guangfu Chemical Co. Ethanol and cyclohexane were both analytical reagents.

The ionic liquids, 1 - butyl 3 - methylimidazole bromide ([bmim]Br) , were synthesized according to the method reported in the literature [14]. SiO2 sols and TiO2 sols were prepared separately. SiO2 sols were generated by mixing TEOS, absolute alcohol, deionized water, glacial acetic acid , urea and

IL ([bmim]Br) together with vigorous stirring in the molar ratio of 1 : 7 : 4.62 : 0.002 : 0.04 : x. TEOS and half of absolute alcohol were fully mixed as solution A and the rest of the materials were fully mixed as solution B, then solution B was added into solution A, drop by drop, while stirring and SiO2 sols were prepared. About 30 minutes later, TiO2 sols were also prepared at the same way by mixing TBT, absolute alcohol, deionized water, glacial acetic acid, urea and IL ([bmim]Br) together in the molar ratio of 1 : 18 : 3.75 : 1.8 : 0.04 : x. Then SiO2 sols and TiO2 sols were mixed together at different SiO2/TiO2 molar ratios and 0.1mol/L aqueous ethanolic silver nitrate was added into the mixed sols after stirring for 10 minutes. The final sols were sharply stirred for about 30 min and then was kept at 60℃ for gelation.

The gels obtained were aged by aqueous IL solution of different constituents at 60℃ for 4 days and then was soaked in excess cyclohexane for 1 day and in excess absolute alcohol for 4 days (change solution four times a day) in order to remove IL in gels. The wet gels were dried at 60℃ for 8 hours and at 80℃ for 8 hours at ambient pressure, and then the aerogels were obtained without calcination.

2.2. Characterization
The aerogel samples were characterized without any other treatment. Nitrogen adsorption-desorption analysis (Nova 1200e Surface Area and Pore Size Analyzer) was used to determine available surface area and pore volume. The average pore size, pore volume and surface area were obtained from BET analysis, applying the cylindrical model. X-Ray diffraction analysis (XRD) was performed using a D/max-3B analyzer (Cu-Ka, 40 kV, 30 mA) to analyze the crystal structure . The surface appearance and porous structure of the aerogels were investigated by electron microscopy (SEM) (Hitachi S-2500).

2.3. Photocatalytic reaction
The photocatalytic performance of the aerogel samples was evaluated by decolorization of organic dye of reactive brilliant blue. The 0.1g block of untreated aerogel samples were used to decolorize 200ml reactive brilliant blue solution (20mg/L). In the first 30 minutes, photocatalytic reaction was performed in the dark. Take the supernatant and measure absorbance by UV spectrophotometer to evaluate adsorption performance. Afterwards, photocatalytic reaction was performed under 145W UV-light irradiation. About 3 hours later, measure the absorbance of the supernatant to evaluate photocatalytic performance of aerogels. The detection wavelength of UV spectrophotometer was performed at 590nm. The photocatalytic degradation rate was calculated by the formula: photocatalytic degradation rate=(A0 — A)/A0*100%.

3. Results and discussion

3.1. Characterization of aerogels
The aerogel samples were prepared at different SiO2/TiO2 molar ratios of 10, 5, 1, 1/5, 1/10 (noted as 10Si-Ti, 5Si-Ti, Si-Ti, Si-5Ti, Si-10Ti) and the x=1. The Ag was doped by Ag/(SiO2 +TiO2)molar ratio of 1% and the aerogels were aged by 1mol/L aqueous IL solution. Obviously, the increase of SiO2/TiO2 molar ratios had been found to prolong the time for gel, increase transparency and reduce cracking of aerogels. The densities of aerogel samples were lower as the increase of SiO2/TiO2 molar ratios.

The porous structures of aerogels were studied by Nitrogen adsorption-desorption. It could be seen from Table 1 that Si-Ti, Si-5Ti and Si-10Ti aerogel samples had significantly higher surface areas than 10Si-Ti and 5Si-Ti aerogel samples. Furthermore, pore volumes of aerogels synthesized with more SiO2 value were lower than pore volumes of aerogels synthesized with more TiO2 value. The Nitrogen adsorption-desorption isotherms of aerogel samples were illustrated in Fig. 1 and isotherm of Si -Ti, Si-5Ti and Si-10Ti aerogels exhibited typical IV curves, which were characteristic of predominantly mesoporous materials based on the IUPAC classification [15]. It was also obvious that isotherm patterns of 10Si-Ti and 5Si-Ti aerogels didn't have the characteristics of any porous microstructure.

These conclusions ran somewhat counter to conventional wisdom and the cause might be that IL made a so great influence on the structural formation of network. Based on the analysis above, the possible mechanisms of IL were conjectured. In the sol-gel process, [bmim]Br played a uniformity and stabilization role to the system due to its special characteristics including formation of high polarizability, hydrogen-bonding networks, low interface tension and low interface energy. It had been known that TEOS had lower hydrolytic condensation rate than TBT. In the preparation process of 10Si-Ti and 5Si-Ti aerogels, since there was not enough TBT to form network structure, small amount of TiO2 particles were synthesized and dispersed in a homogeneous phase due to special solubility and high polarizability of IL. On account of lots of TEOS, huge SiO2 network structure was formed. TiO2 particles were locked into in the pores and SiO2 network structure and this led to the very small surface area of 10Si-Ti and 5Si-Ti aerogels. In the preparation process of Si-5Ti and Si-10Ti aerogels, there was lots of TBT in the homogeneous system so that TiO2 network structure was formed.

Simultaneously, little TEOS polymerized with TBT to form network structure but not formed particles, which prevented blocking network structure, so Si-5Ti and Si-10Ti aerogels had a high surface area. In the preparation process of Si-Ti, since the same amount of TEOS and TBT were dispersed in a homogeneous phase, SiO2-TiO2 hybrid network was formed and this made the highest surface area of Si-Ti aerogels. In consideration of lower hydrolytic condensation rate of TEOS than TBT, there was SiO2 network mostly in the outermost of aerogels so that it was very rough and uneven in surface of aerogels, matching the conclusions from SEM morphology in the sections below.

Table 1. BET analysis results of the aerogel samples at different SiO2/TiO2 molar ratios

sample	10Si-Ti	5Si-Ti	Si-Ti	Si-5Ti	Si-10Ti
Surface Area/m^2g^{-1}	95.6	93.6	484.9	396.8	385.2
Pore Radius/nm	3.948	4.126	11.442	10.322	5.296
Pore Volume/cm^3g^{-1}	0.025	0.024	0.776	0.758	0.549

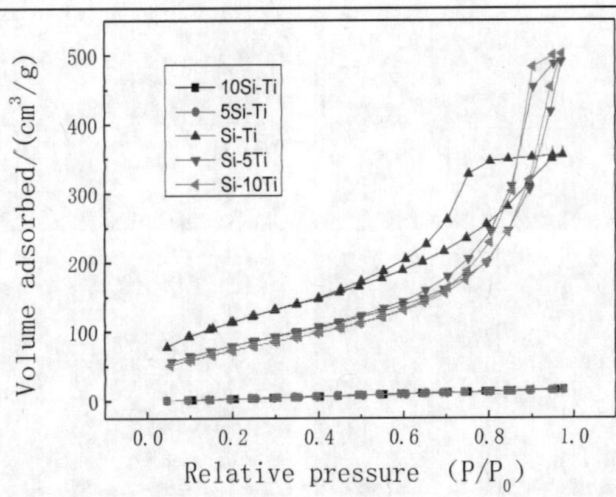

Figure 1. Nitrogen adsorption-desorption isotherms of the aerogels made in different SiO2/TiO2 molar ratios (10Si-Ti, 5Si-Ti, Si-Ti, Si-5Ti, Si-10Ti). 1% Ag was doped in aerogels and IL was used by n(IL)/n(SiO2+TiO2) =1

The X-ray powder diffraction (XRD) patterns of aerogel samples were shown in Fig.2 and it showed that the different SiO2/TiO2 molar ratios had a significant impact on the formation of anatase phase of the TiO2. Fig.2 indicated that IL ([bmim]Br) had the capacity for good anatase crystallinity under the premise of enough TiO2 in the SiO2/TiO2 aerogels even though aerogels were prepared by ambient pressure drying at low temperature (60℃and 80℃) and without calcination. The crystalline nature became more prominent with increase in the content of TiO2 in SiO2/TiO2 aerogels. The pattern of Si-10Ti aerogel sample showed the diffraction peaks assigned to anatase structure of TiO2 and the crystalline phase was indexed in the light of available XRD data for TiO2 crystal structure. The size of crystal was roughly calculated by Scherrer Equation. Increased SiO2/TiO2 molar ratios

IOP Conf. Series: Materials Science and Engineering **587** (2019) 012016 doi:10.1088/1757-899X/587/1/012016

resulted in bigger anatase nanocrystals and average sizes ranged from 0.4nm for 10Si-Ti to 5.2 nm for Si-10Ti.

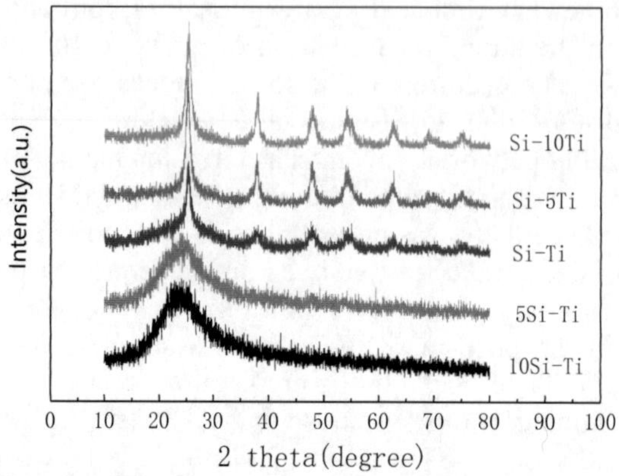

Figure 2. X-ray powder diffraction (XRD) patterns of aerogel samples made in different SiO2/TiO2 molar ratios (10Si-Ti, 5Si-Ti, Si-Ti, Si-5Ti, Si-10Ti).

Fig.3 showed the SEM morphology of composite aerogels(10Si-Ti, Si-Ti, Si-10Ti) and pure TiO2, respectively. The results stated that aerogels except the aerogel synthetized in SiO2/TiO2 molar ratio of 10 had nano-porous microstructure. From Fig.3 (a), the 10Si-Ti aerogel was composed of dense layers and did not show any porous microstructure under SEM observation. From Fig.4 (c), it indicated clearly that Si-10Ti aerogels were densely agglomerated by nanoparticles and of porous structure. Pure TiO2 aerogels had a fundamentally similar image to Si-10Ti aerogels. In the Fig.4 (b), SEM image showed that Si-Ti aerogels had a looser network structure and larger pores. These results were consentaneous with table1 and confirmed the conjecture about the possible mechanisms of IL.

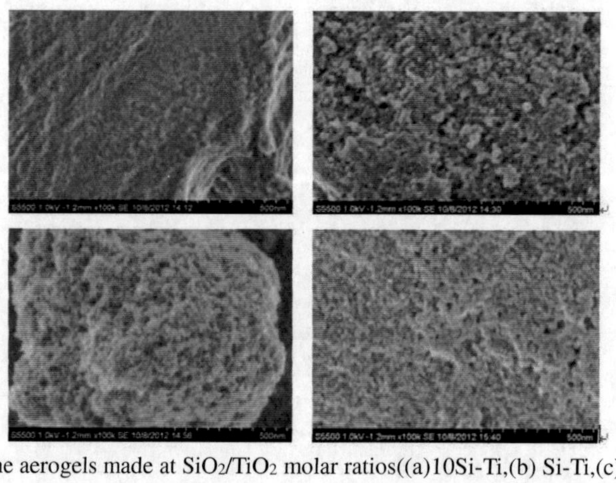

Figure 3. SEM images of the aerogels made at SiO_2/TiO_2 molar ratios((a)10Si-Ti,(b) Si-Ti,(c) Si-10Ti) and(d) Pure TiO_2

3.2. Photo-catalytic performance of aerogels

It was showed in the Fig.4 that photocatalytic degradation rate of the aerogel samples was increased with the decrease of SiO2/TiO2 molar ratios regularly. Degradation reaction was carried out without the UV light in the dark at the first 30 minutes and reactive brilliant blue was largely adsorbed by aerogels because of its absorption capability. It was seen from Fig.4 that the degradation rates of the first 30 minutes corresponded to the surface areas of aerogel samples and the conclusion illustrated the surface area of aerogel played an important part in the photocatalytic performance. Later on, degradation reaction continued for 150 minutes under single UV irradiation and anatase TiO2 played a greater role in photocatalytic reaction. The photocatalytic degradation rate of the aerogel sample in Si-10Ti was best up to 99.77% at 170 minutes. The result shows photocatalytic performance was both relevant to high surface area and anatase micro-crystal phase. After photocatalytic reaction, most of the aerogel samples were still blocky so that aerogels as catalyst would been more easily recycled.

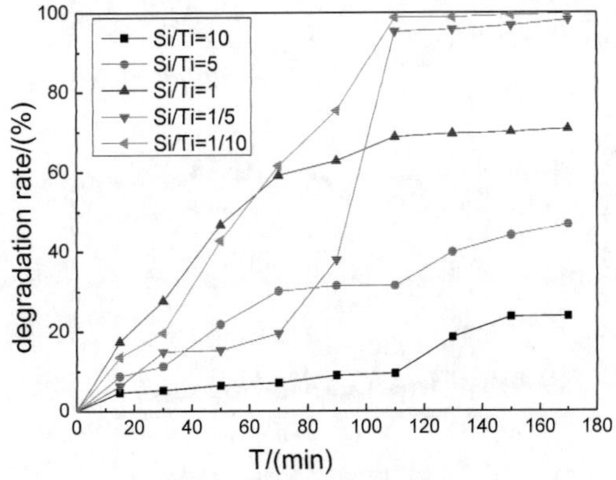

Figure 4. Adsorption and photocatalytic degradation rate curves for 200 mL 20 mg/L reactive brilliant blue solution with 0.1g blocky aerogel samples made at different SiO_2/TiO_2 molar ratios (10Si-Ti, 5Si-Ti, Si-Ti, Si-5Ti, Si-10Ti).

3.3. Effect of doping Ag nanoparticles

The aerogels were prepared in different ratios n(Ag)/n(SiO2+TiO2)=0, 1%, 5%, 10% (noted as 0Ag-Si-10Ti, 1Ag-Si-10Ti, 5Ag-Si-10Ti, 10Ag-Si-10Ti,) in order to investigate the effect of various Ag values to aerogels. The SiO2/TiO2 molar ratio of aerogels was fixed at 1/10 and the IL/(SiO2+TiO2) molar ratio was fixed at 1(x=1). The aerogels were aged by 1mol/L IL aqueous solution.

The XRD patterns of Ag doped SiO2-TiO2 aerogels were displayed in Fig.5. It was obviously seen that the pattern of 10Ag-Si-10Ti aerogel sample showed the diffraction peaks which the XRD patterns of other aerogel samples didn't have and the peaks were in agreement with peak position of silver metal. The diffraction peaks were corresponded to the (110) planes of silver shifted from 39° to 31° . The cause of the disappearance of other patterns might be that there was too little Ag value to detect.

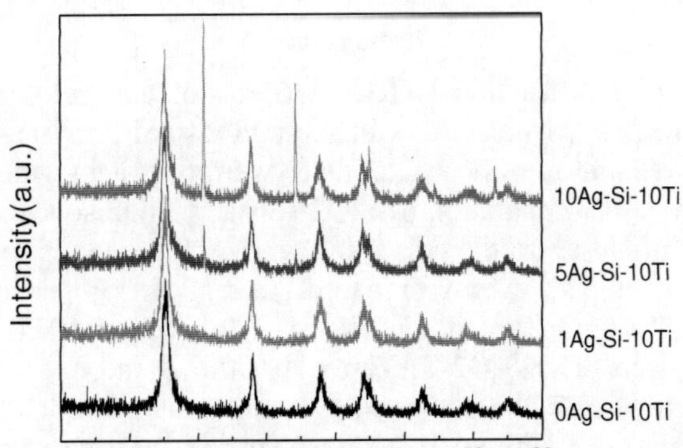

Figure 5. X-ray powder diffraction (XRD) patterns of aerogel samples with various Ag values

The values of Ag doped aerogels also made an important effect on the photocatalytic performance [16-19]. Both 0Ag-Si-10Ti and 10Ag-Si-10Ti aerogel samples (0.1g) were used to decolorize 200ml reactive brilliant blue solution (20mg/L) under UV-light. After 1 hour later, the photocatalytic degradation rate of 10Ag-Si-10Ti aerogel sample was up to 88.61% while the photocatalytic degradation rate of 0Ag-Si-10Ti aerogel sample was only 61.54%. The result showed Ag could improve photocatalytic activities.

3.4. Effect of IL on anatase crystallinity

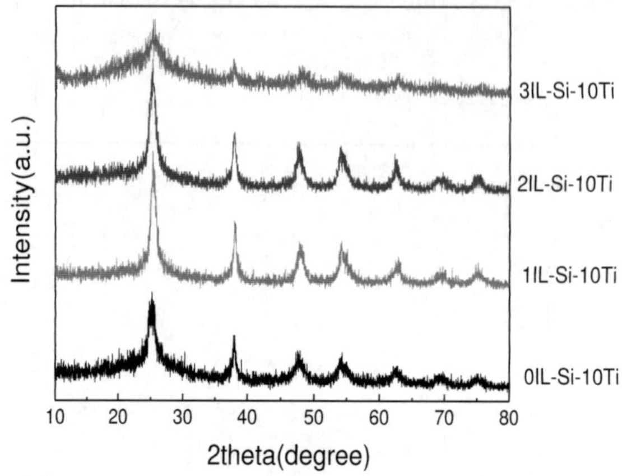

Figure 6. (a) X-ray powder diffraction (XRD) patterns of aerogel samples prepared at different ratios n(IL)/n(SiO2+TiO2)=0, 1, 2, 3 in the first step(0IL-Si-10Ti, 1IL-Si-10Ti, 2IL-Si-10Ti, 3IL-Si-10Ti)

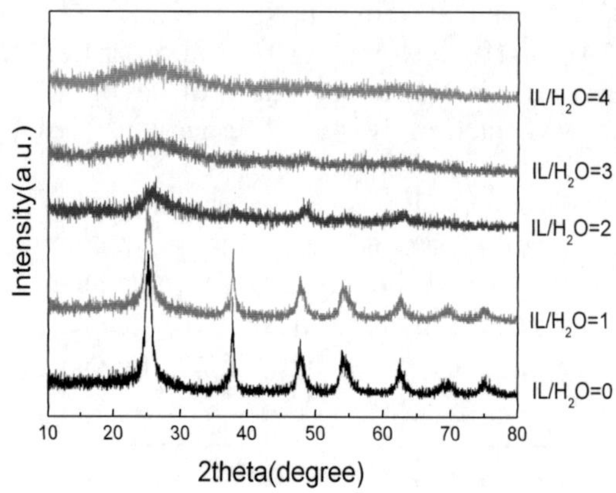

Figure 6. (b) X-ray powder diffraction (XRD) patterns of aerogel samples aged by variable IL aqueous solution of 0, 1mol/L, 2mol/L, 3mol/L, 4mol/L in the second step

In the process of the preparation of aerogels, there were two steps where IL ([bmim]Br) was used as follows: the first step was the prepatation of SiO2 sols and TiO2 sols and IL was used as cosolvent; the second step was aging process and the gels were aged by IL aqueous solution.

Aerogels without Ag doped were prepared using IL as cosolvent at different ratios n(IL)/n(SiO2+TiO2) =0, 1, 2, 3 (noted as 0IL-Si-10Ti, 1IL-Si-10Ti, 2IL-Si-10Ti, 3IL-Si-10Ti, respectively) in the first step. The SiO2/TiO2 molar ratio of aerogels was fixed at 1/10 and the aerogels were aged by deionized water rather than IL aqueous solution. The other series of aerogels without Ag doped were prepared without IL as cosolvent in the first step and the SiO2/TiO2 molar ratio of aerogels was also fixed at 1/10. The aerogels were aged by variable IL aqueous solution of 0, 1mol/L, 2mol/L, 3mol/L, 4mol/L (noted as Si-Ti-0, Si-Ti-1, Si-Ti-2, Si-Ti-3, Si-Ti-4, respectively).

These aerogels were analyzed by XRD and the size of crystal was roughly calculated by Scherrer Equation. The patterns in Fig.6(a) showed the aerogels using IL as cosolvent appeared microcrystalline anatase and it was obviously that [bmim]Br played a very important role in the size of crystal. 1IL-Si-10Ti and 2IL-Si-10Ti aerogels had bigger crystal grain dimension of 6.2nm and 5.6nm, respectively, than 0IL-Si-10Ti aerogels of 2.6nm,and on the contrary, 3IL-Si-10Ti aerogels had smaller crystal grain dimension of 1.7nm. The patterns in Fig.6 (b) showed the aerogels, which aged with deionized water instead of IL aqueous solution, also appeared microcrystalline anatase. Si-10Ti-1 and Si-10Ti-2 aerogels had bigger crystal grain dimension of 6.6nm and 5.3nm than 0IL-Si-10Ti aerogels of 2.6nm while Si-10Ti-3 and Si-10Ti-4 aerogels didn't have TiO2 microcrystalline anatase nearly. These showed that appropriate [bmim]Br used as cosolvent and IL aqueous solution of proper concentration used as aging agent were both beneficial to form microcrystalline anatase. Under optimal conditions of n(IL)/n(SiO2+TiO2)=1 and IL aqueous solution of 4mol/L, the best anatase crystallinity were achieved to 6.2nm and 6.6nm, respectively.

Special solubility and dispersing effect of IL to sol–gel precursors were estimated to cause the results above. Appropriate IL made sol–gel precursors distributed evenly in a homogeneous phase, which had a positive effect on formation of microcrystalline anatase with larger crystal size. Whereas, excess IL engendered too high mucosity of solution and this was disadvantageous for the diffusion and polycondensation of sol–gel precursors. In this case, it was difficult to form microcrystalline anatase of the big size, and even anatase.

4. Conclusions

Ag doped SiO2-TiO2 aerogels with nano-sized microcrystalline anatase and high surface area were successfully prepared using IL ([bmim]Br) by ambient pressure drying at low temperatures and without calcination. It was found that the SiO2/TiO2 molar ratios influenced aerogels formation and the optimal SiO2/TiO2 molar ratio was 1/10. Aerogels in SiO2/TiO2 molar ratio of 1/10 had a surface area as high as 385.2m2g-1, big anatase crystallinity of 6.2nm and good photocatalytic degradation rate up to 99.7%. It was also found that the increase of Ag amount would effectively improve photocatalytic abilities of aerogels. Appropriate [bmim]Br used as cosolvent and IL aqueous solution of proper concentration used as aging agent were both beneficial to form microcrystalline anatase. In the paper, the optimal conditions were n(IL)/n(SiO2+TiO2)=1 and IL aqueous solution of 1mol/L. However, mechanisms of IL was not entirely clear and needed to work on later.

References

[1] L. W. Hrbesh, Chem. Ind. 24 (1990) 824.
[2] A. C. Pierre, G. M. Pajonk, Chem. Rev. 102 (2002) 4243.
[3] J. Fricke, T. Tillotson. Thin Solid Films. 297 (1997) 212.
[4] Y. K. Akimov, Instrum. Exp Tech. Rev. 46 (2003) 287.
[5] L. W. Hrubesh, J. Non-Cryst. Solids. 225 (1998) 335-342.
[6] J. X. Liu, X.W. Ling, F. Shi, X. L. Dong, B. Zhai, Y. G. Li, J. Chin. Ceramic Soc. 38 (2010) 2296.
[7] M. A. Klingshirn, S. K. Spear, J. D. Holbrey, R. D.Rogers, J. Mater. Chem. 15 (2005) 5174.
[8] A. Karout, A. C. Pierre, Catal. Commun. 10 (2009) 359.
[9] P. Y. Yu, Y. P. Zhang, J. X. Zhang, X. D. Yang, Chem. Ind. & Eng. Pro. 29 (2010) 1251.
[10] S. Dai, Y. H. Ju , H. J.Gao , J. S. Lin , S. J. Pennycook, C. E. Barnes, Chem. Commun. 3(2000) 243.
[11] K. Yoo, H. Choi, D. D. Dionysiou, Catal. Commun. 6 (2005) 259.
[12] F. Zhang, D. Sun, C. Yu, Y.Yin, H. Dai.. G Shao, New J. Chem. 39(2015)3065.
[13]J T. Han, X.Wang,Y Ma, G. Shao, X Dong,C Yu. Sol-Gel Sci Technol. 82 (2017):101.
[14] J. G. Huddleston, A. E. Visser, W. M. Reichert, H. D. Willauer, G. A. Broker, R. D. Rogers, Green Chem. 3 (2001) 156.
[15] T. J. Barton, L. M. Bull, W. G. Klemperer, Chem. Mater. 11 (1999) 2633.
[16]C.Liu, D.Yang, Y. Jiao, Y. Tian, Y. Wang, Z. Jiang ACS Appl. Mater. Interfaces, 5 (2013) 3824.
[17] S. K. Cushing, J.Li, J. Bright, B.T. Yost, P. Zheng, A.D. Bristow, N. Wu J. Phys. Chem. C, 119(2015) 16239.
[18] F.Wang, J.Ma, G.He, M.Chen, S.Wang, C. Zhang, H. He Ind. Eng. Chem. Res., 57(2018) 11903.
[19]Y.Huang, Q.Yu, J.Wang, J.Wang, C.Yu, J.T.Abdalla, Z.Zeng, S.Jiao, D.Wang, S. Gao ACS Sustainable Chem. Eng. 6(2018) 438.

IOP Conf. Series: Materials Science and Engineering **587** (2019) 012017 doi:10.1088/1757-899X/587/1/012017

Fracture failure analysis of driving crankshaft of substation circuit breaker

D K Li[1*], X Liu[2], M C Duan[3], J W Yang[4] and Y Xie[1]

[1]State Grid Hunan Electric Power Corporation Research Institute, Changsha, Hunan, 410007, China
[2]State Grid Huaihua Power Supply Company, Huaihua, Hunan, 418000, China
[3]State Grid Chenzhou Power Supply Company, Chenzhou, Hunan, 423000, China
[4]State Grid Xiangtan Power Supply Company, Xiangtan, Hunan, 411104, China

Corresponding author and e-mail: D K Li, 118095657@qq.com

Abstract. Fracture failure analysis of the driving crankshaft of a substation circuit breaker was conducted by means of macroscopic inspection, magnetic particle testing, spectroscopic analysis, mechanical property test and metallographic examination. The analysis results show that the fractured crankshaft is in the cast state, the manufacturer has not carried out follow-up heat treatment after casting, and there are large residual stresses and even cracks in the castings. Finite element analysis shows that the new driving crankshaft produced by Q235-A forged steel plate has higher safety performance and can replace the driving crankshaft produced by casting process.

1. Introduction

The opening and closing operation of circuit breaker is realized by the contact and separation between the moving contact and the static contact, and the movement of the moving contact is driven by the crankshaft. Therefore, the role of the crankshaft in the circuit breaker is very important. The quality of the crankshaft will directly affect the accuracy of the circuit breaker opening and closing. If the crankshaft breaks suddenly during operation, it will lead to unsuccessful opening and closing of circuit breaker, and will cause major safety accidents [1].

Driving crankshaft of the 508 circuit breaker broke during the pre-commissioning test of a 110 kV transformer substation. The circuit breaker was operated only less than 100 times before the fracture. The circuit breaker manufacturer is Shandong Tai'an Taishan High Voltage Switch Co., Ltd. The spring mechanism model of the operating mechanism is CTB-I, and the design material of the crankshaft is ZG 310-570. In this paper, the fracture reasons of the driving crankshaft are analyzed and the improvement measures are put forward.

2. Fracture Cause Analysis

2.1. Macroscopic inspection

The driving crankshaft of the circuit breaker spring mechanism is completely broken, as shown in figure 1. The crack zone with obvious rust and the silver-white final fracture zone can be seen on the fracture surface. The crack zone almost runs through the whole cross section, as shown in figure 2, which means that there are serious cracks in this crankshaft before the field test. Thus, the driving crankshaft suddenly broke after only a few dozen actions.

Visual inspection of the fractured crankshaft and the same batch of crankshaft shows that there are rust and holes on the inner surface, as shown in figure 3, which means that there are some problems in the manufacturing process of this batch of crankshaft.

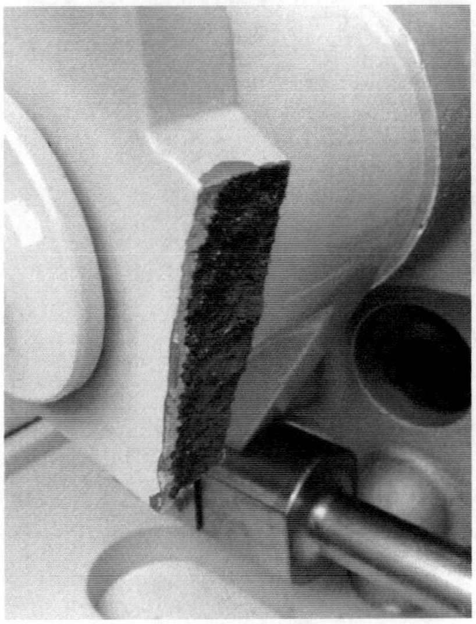

Figure 1. Fracture of the driving crankshaft.

Figure 2. Macroscopic fracture morphology.

Figure 3. Rust and holes on the inner surface.

2.2. Magnetic particle testing

Two driving crankshafts in the same batch were sampled and the paint film was polished off for magnetic particle testing. It was found that there were dense cracks in both crankshaft. The maximum crack length was about 20 mm, as shown in figure 4. According to GB/T 9444-2007 "Magnetic Particle Detection of Steel Castings", the quality grade is determined as LM5, which means Linear Defect Grade 5. Therefore, the manufacturing quality of this batch of driving crankshafts is not up to standard.

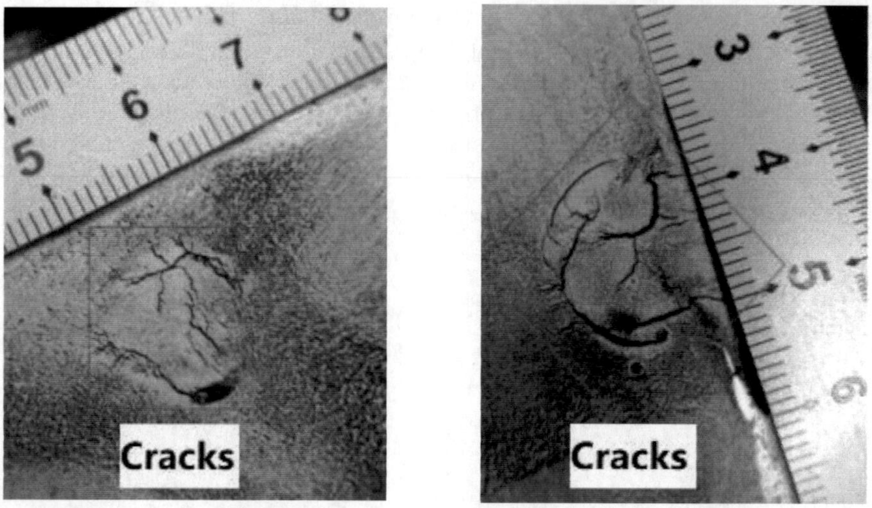

Figure 4. Cracks on the crankshaft surface revealed by magnetic particle inspection.

2.3. Spectroscopic analysis

Spectroscopic analysis was carried out on the fractured crankshaft. The experimental results are shown in Table 1, and the composition of the cast steel brand ZG 310-570 in GB/T 11352-2007 standard is also listed for comparison. It can be concluded that the chemical composition of the fractured crankshaft meets the standard requirements.

Table 1. Chemical composition analysis of the fractured crankshaft.

Samples	Chemical Composition/ %				
	C	Si	Mn	P	S
The fractured crankshaft	0.210	0.295	0.505	0.0176	0.0125
ZG 310-570	≤0.50	≤0.60	≤0.90	≤0.035	≤0.035

2.4. Mechanical property test

Room temperature mechanical properties of the fractured crankshaft was carried out. The experimental results are shown in Table 2, and the mechanical properties of the cast steel brand ZG 310-570 in GB/T 11352-2007 standard is also listed for comparison. It is found that the prescribed plastic elongation strength $R_{p0.2}$ and tensile strength R_m of the fractured crankshaft are far greater than the lower limit value of ZG 310-570 in the standard, while the elongation after fracture is significantly lower than the standard value. Therefore, it is concluded that the room temperature mechanical properties of the fractured crankshaft are abnormal.

Table 2. Mechanical properties of the fractured crankshaft.

Samples	Mechanical Properties Parameters		
	$R_{p0.2}$/MPa	R_m/MPa	A/%
The fractured crankshaft	720	992	7.7
ZG 310-570	⩾310	⩾570	⩾15

2.5. Metallographic examination

The metallographic structure of the fractured crankshaft is shown in figure 5. It is composed of Widmanstatten structure of fine strip ferrite, massive ferrite and pearlite, showing typical metallographic morphology of ZG 310-570 castings [2].

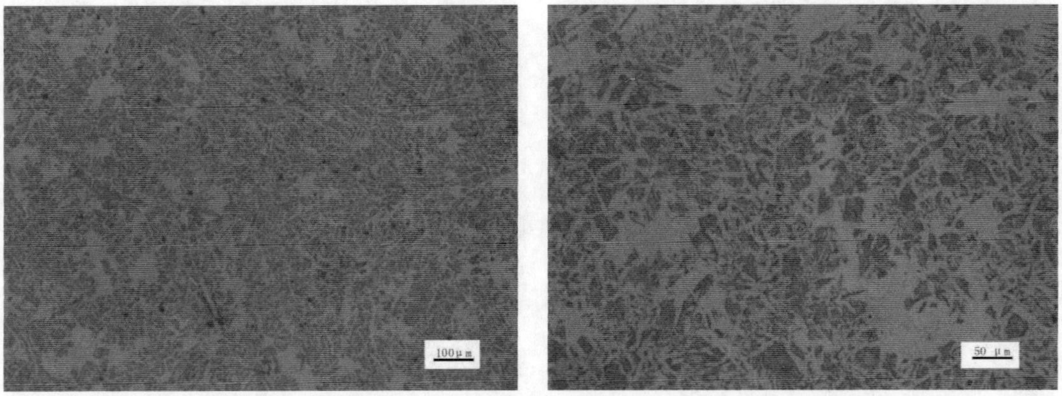

Figure 5. Metallographic structure of the fractured crankshaft.

ZG310-570 cast steel has medium carbon content, and it is easy to form Widmanstatten structure in as-cast state, which results in poor mechanical properties. Heat treatment can improve its metallographic structure and comprehensive mechanical properties. Heat treatment can also eliminate the residual stress of the castings and avoid deformation and cracks in the process of using the castings. ZG 310-570 steel castings are usually heat treated by annealing, normalizing or normalizing plus tempering. The heat treatment specification is to heat the furnace to 900±20°C and cool it after holding for 4 hours. The cooling mode can choose furnace cooling or stacking air cooling according to site specific conditions and operation requirements, which are also known as annealing and stacking normalizing, respectively [3]. Because the internal stress of normalized castings is larger than that of annealed castings, the heat treatment process of tempering after normalizing can be adopted for some important parts. The normal normalizing structure of ZG310-570 cast steel is white fine ferrite with uniform distribution and black fine pearlite, so that the casting has the best comprehensive mechanical properties [3].

Therefore, it can be judged that the metallographic structure of the fractured crankshaft is not qualified. Based on the above mechanical properties test results, it can be concluded that the basic reason for the unqualified structure and abnormal mechanical properties of the castings is that the castings are not annealed, normalized or normalized plus tempered. That is to say, the fractured crankshaft is directly put into use without any heat treatment after casting.

The above analysis shows that the fundamental reason for the fracture of the driving crankshaft is that the crankshaft in this batch is in as-cast state. The manufacturer has not carried out follow-up heat treatment for the castings, and there are large residual stresses or even cracks in the castings, which can not meet the quality standards of the castings [4].

3. Improvement measures

Fracture analysis of driving crankshaft shows that the quality of this batch of crankshaft manufactured by casting method is not up to standard, and the long-term safe operation can not be guaranteed. After communication and coordination with the manufacturer, it is proposed to replace all the six driving crankshaft of the substation in the same batch. The material of the new driving crankshaft is Q235-A steel plate, and the manufacturing process is forging.

In order to ensure the quality of the new driving crankshaft, one is randomly selected for experimental testing. Tensile tests at room temperature show that the lower yield strength, tensile strength and elongation after fracture of the new crankshaft are 322 MPa, 452 MPa and 40.3%

IOP Conf. Series: Materials Science and Engineering **587** (2019) 012017 doi:10.1088/1757-899X/587/1/012017

respectively. The metallographic structure of the new crankshaft is shown in figure 6, showing a streamlined shape, indicating that it is a typical forging structure[5].

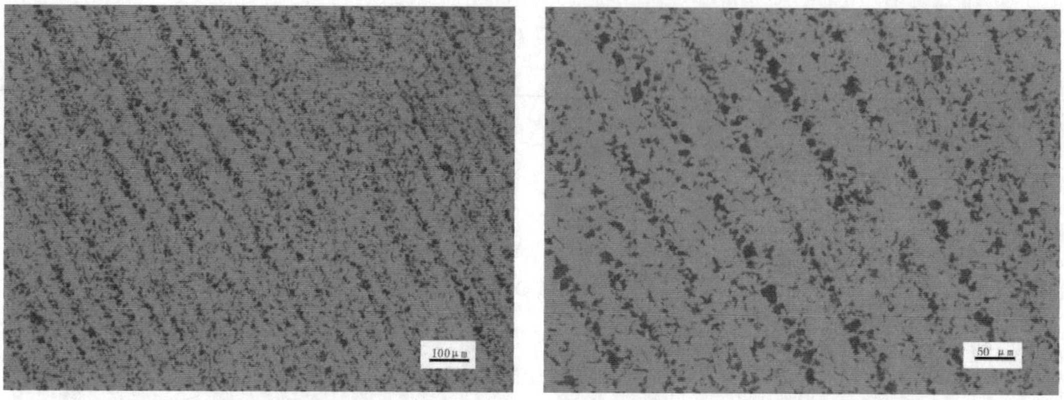

Figure 6. Metallographic structure of the new crankshaft.

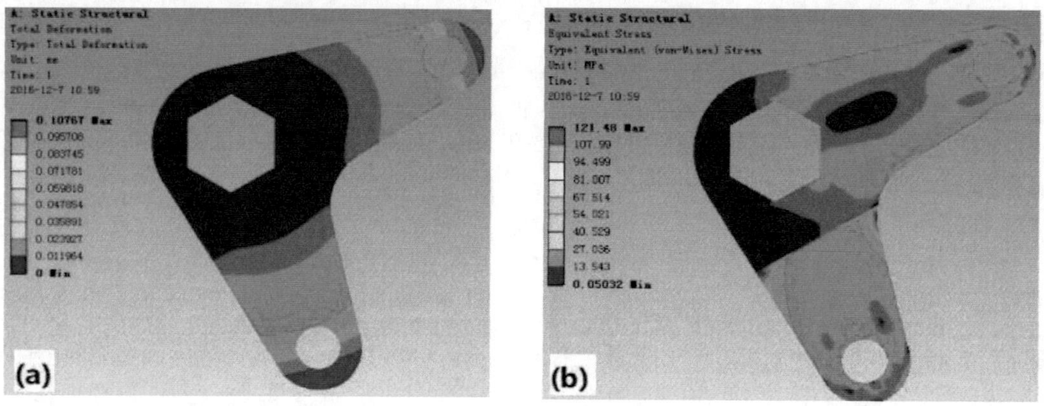

Figure 7. Finite element calculation of the new crankshaft (a) total deformation (b) equivalent stress.

The finite element method is used to calculate the stress and deformation of the new crankshaft, as shown in figure 7. The calculation results show that under service conditions, without considering the stress concentration, the deformation is 0.1 mm and the stress is 108 MPa. If the yield strength of the material is 322 MPa, the safety factor is 3. Therefore, the mechanical properties of the new crankshaft can fully meet the application requirements.

4. Conclusions

Reason for fracture of the driving crankshaft of the circuit breaker in this substation is that the crankshaft in this batch is in as-cast state, the manufacturer has not carried out follow-up heat treatment for the castings, and there are large residual stresses or even cracks in the castings, which can not meet the quality standards. The new driving crankshaft produced by Q235-A forged steel plate has higher safety performance and can replace the driving crankshaft produced by casting process.

The driving crankshaft of circuit breaker bears large impact load during its service. If a sudden fracture occurs, it will lead to serious electrical equipment accidents. In view of the better comprehensive mechanical properties and better safety performance of forged steel parts than cast steel parts, it is suggested that forgings be used for important load-bearing parts in substation, such as driving crankshaft of circuit breaker, operating abutment and driving connecting rod, etc.

IOP Conf. Series: Materials Science and Engineering **587** (2019) 012017 doi:10.1088/1757-899X/587/1/012017

References

[1] Wang Z, Zhu Q, Rao B Q 2014 *East China Electric Power* 12: 2641-2644. (In Chinese)
[2] Lu W B, Chen R F, Zhang A P 2016 *Castings* 02: 31-34. (In Chinese)
[3] Wang Z M 2017 *Foundry Technology* 04: 843-847. (In Chinese)
[4] Wang J W, He J, Zhao T, et al. 2015 *High Voltage Electrical Appliances* 04: 200-204. (In Chinese)
[5] Zhang X Y, Xu X J, Chen R F, et al. 2018 *Forging Technology* 01: 25-30. (In Chinese)

Institute of Physics
Dirac House, Temple Back
Bristol BS1 6BE UK

ISSN: 1757-8981
ISBN 978-1-5108-9460-0

CURRAN ASSOCIATES INC.
proceedings
.com